PLANET OF THE TREES

The human race and its climate gamble

JOHN HALKETT

Preface by John Hewson

HALSTEAD PRESS

CANBERRA MMXXI

Published by Halstead Press
Gorman House, Ainslie Avenue
Braddon, Australian Capital Territory, 2612

and

Unit 66, 89 Jones Street
Ultimo, New South Wales, 2007

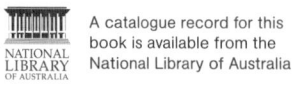 A catalogue record for this book is available from the National Library of Australia

The paper for the pages of this book is from a mill certified to have used material from responsible sources. No old growth timber has been used in its manufacture.

ISBN: 978 1925043 648

All known images sourced from Shutterstock, unless otherwise indicated.

CONTENTS

PREFACE

by John Hewson

A book that could perhaps be described as a post-apocalyptic narrative. However, as science and recent history tell us, unless we are much more affirmative, and quickly, about addressing climate change, humanity's future on this planet is under real and direct threat.

This book builds on Halkett's last work, *By the Light of the Sun: Trees, Wood, Photosynthesis and Climate Change*, that details how harnessing the power of the products of photosynthesis could assist in offsetting adverse climate change. Specifically, the book asserts that trees and forests will be critical ingredients in the search for a zero net carbon emissions future.

I make no apologies for being vigorous in my climate change advocacy, and do not resile from my criticism of inept government policy when it comes to tackling climate change. I will certainly continue being a strong and persistent voice for desperately needed action.

While Australia's Prime Minister Scott Morrison has said publicly that he does not deny that the climate is changing, nor that the burning of fossil fuels is the primary cause, he persistently will not admit to the severity of the crisis, nor the reality that his government is failing to respond adequately.

To our great shame as Australians, our international reputation took a severe hammering over the 2019/20 summer as the world's media was full of apocalyptic images of kilometre high flames, exploding forests, fleeing kangaroos and burnt koalas. However, unlike the rest of the world, the Australian Government refused to link the fire atrocity to climate change.

Australia that once prided itself on being a good international citizen, when it came to matters of global importance, has now tragically witnessed a decade or more of poor national leadership on climate change, with a succession of prime ministers lacking either the intellect or courage to develop coherent policy responses.

The sorry litany of failed attempts is on the record. Prime Minister Rudd's squabbling about the "greatest moral challenge of our time". Prime Minister Gillard's bumbling over whether or not an emissions reduction scheme was, or was not, a tax. Prime Minister Abbott's ruthless exploitation of this blunder, and his 'weaponising' of climate change. Then there was Prime Minister Turnbull's inability to face down the climate change deniers in his party.

Yet successive prime ministers have been increasingly aware that the smart money was moving away from fossil fuels, notably coal, and of new opportunities for climate change-friendly innovation and investment.

Capital is deserting fossil fuel projects, because renewables are an increasingly attractive investment proposition and community campaigns pressure banks and investment funds to stop investing in fossil fuel-based projects.

Some banks have stopped, or are proposing to stop lending to fossil fuel projects. Some insurers are not insuring them, and big asset owners, including sovereign wealth funds and super funds, are exiting climate exposed investments. The Adani corporation has so far been unable to find a bank willing to fund its huge Carmichael mine in Queensland's Galilee Basin, and three of Australia's big four banks have pledged to stop lending to thermal coal projects.

Major companies including BHP and Qantas, as well as the Minerals Council of Australia, the Australian Petroleum Production and Exploration Association, the National Farmers Federation, and the Business Council of Australia have rolled out climate change strategies in the past two to three years. The Australian Medical Association has formally declared climate change a health emergency, pointing to "clear scientific evidence indicating severe impacts for our patients and communities now and into the future".

In October 2020, a global push for net zero emissions by investors, corporations and regulators convinced the ANZ Bank to overhaul its carbon policies, laying the foundation for sustainable finance opportunities and a recognition, as one of the world's top banks, of the merits of helping industries transition to a low carbon economy.

ANZ points to the huge growth in green finance—a global market worth about $793 billion, that includes opportunities in energy efficiency, low emissions transport, green buildings, reforestation, renewable energy and battery storage, emerging technologies such as carbon capture and storage and hydrogen-based technology, disaster resilience and climate change adaptation measures.

ANZ has announced that it will implement an ambitious net zero emissions action plan and incorporate climate change in its condition of lending—increasing pressure on farmers, construction firms and a range of companies to establish low carbon transition plans in 2021 as it ramps up support for the Paris Agreement goal of net zero emissions by 2050. The Bank will also move away from working with customers that don't have clear and public climate change transition plans, and by 2025 will fund and facilitate at least $50 billion annually to support sustainable solutions to help customers reduce carbon emissions.

In addition, we know that community activism is on the rise. Since 2010 an alliance of environmental, community and farming groups calling itself Lock the Gate has been opposing coal and coal seam gas developments. The mission of this alliance is: "To protect Australia's natural, cultural and agricultural resources from inappropriate mining and to educate and empower all Australians to demand sustainable solutions to food and energy production". The campaign calls on landowners to "lock the gate" to prevent access to the land by gas companies, and to refuse to negotiate the sale of properties to coal mining companies.

In dealing with the crisis of the corona virus pandemic, to its credit Australia has taken an evidence-based bipartisan policy approach. Politicians listened to scientists, and on their advice did not hesitate to inflict severe economic hardship. A Liberal Prime Minister worked effectively with Labor and Liberal state and territory premiers and ministers. They put ideology and vested interests aside and achieved remarkable results, protecting the country from the trauma that the pandemic has inflicted on other countries.

With the pandemic having paused so much of the world's economy, Australia is arguably at a crossroads. We have the chance to shake off the curse of fossil fuels, to prevent the catastrophic fires and heatwaves scientists have long predicted, the species extinctions and the famines.

The pandemic should be seen as a dress rehearsal for what awaits us if we continue to ignore the laws of science, the physical world, and the demands of catastrophic climate change threats. Existing systems and practices that have been taken for granted—our economic system, our food system, our energy system, our transport system, our production and waste systems, our governance mechanisms, our lifestyles and our relationship with nature—must undergo searching examination and reform if humanity is to survive.

The pandemic has been an inspiring indication that the community can embrace difficult but essential change. The risks emerging from the corona virus were varied, complex, global and catastrophic, and the solutions needed to be national, globally collaborative and multi-disciplinary. The corona virus will be overcome, but climate change will remain as an existential threat to humanity.

However, when it comes to climate change, the Australian government seems incapable of accepting scientific and other evidence, and fails to listen to clear warnings and predictions. It also appears unwilling to plan for how to avoid or manage mounting catastrophic risks that are threatening our living standards and lifestyle, and in the end the very survival of humanity.

A declared focus of the Morrison Government during the corona virus pandemic was to "keep Australians safe and secure". Surely climate change is now our major national security issue, out-ranking the risk of invasion, terrorism and regional insecurity.

Unfortunately, an entrenched, anti-climate sentiment in the government remains. This was exemplified by comments made by Prime Minister Morrison at the United Nations[1] that Australia is going to meet its emissions targets—a gross misrepresentation which is staggering for someone in his position.

The Prime Minister does not accept either the magnitude or the urgency of the climate challenge, and is beholden to the fossil fuel lobby. Several of his senior staff are ex-coal executives; a couple of his key ministers have coal industry links, and fossil fuel companies are major donors to political parties.

A key question is just how long the prime minister can sustain his defence of the indefensible. To buy time he is attempting to do as little as possible. He hopes voters' memories of his incompetence and lack of empathy will fade. Similarly, his talk of state responsibilities and federal-state tensions and constitutional hurdles is a diversion, an attempt to shift blame.

Alternatively, Prime Minister Morrison could, for example, develop a national program of regenerative agriculture, to make our soils more resilient and drought resistant. And he could certainly develop transition pathways for key sectors to accelerate emissions reductions. Or he could, as does this book, put a spotlight on the role of trees in contributing to improved climate change abatement outcomes.

It is crystal clear that extreme weather events could devastate large parts of Australia and radically impair food production, water availability, public health, infrastructure, the community and the financial system. Comprehensive research[2] has found Australians are increasingly concerned about droughts and floods, extinctions and water shortages associated with climate change, and most people think all levels of government are just not doing enough to combat the effects of global warming.

In spite of a community consensus there is still a great degree of reluctance amongst members of the Australian Government to initiate meaningful action. They are still resisting a commitment to net zero emissions, in spite of the fact that many countries have done so.

We need a commitment to end fossil fuel subsidies; we need a commitment to accelerate the closure of coal-fired power stations and to oppose new gas generation. Again, the government must back renewable energy solutions instead of ignoring and neglecting signals from investors, climate scientists and the electorate calling time on fossil fuel projects.

I think the ability to see that there are positive solutions, and to be able to think about how we might convert something that sounds like all doom-and-gloom has been lacking.

Under President Joe Biden the United States of America is expected to go down a pathway of a rapid transition to zero emissions energy, and may introduce carbon border tariffs, which, along with the work of the European Union, could transform the level of international pressure on countries like Australia that are seen as laggards on climate action.

President Biden has already committed to a green energy program, expected to cost about $US4 trillion over four years. He has undertaken to achieve net zero emissions by 2050, and to decarbonise electricity by 2035.

If humanity is to overcome the risk outlined in this book of following dinosaurs, neanderthals and other life forms into extinction, urgent action is needed. A willingness to undertake such action has not been sufficiently acknowledged and implemented in Australia to date. Consensus and strong leadership are needed if the long-term survival prospects for humanity are to be elevated from current improbable levels.

PREAMBLE

Australia today is ground zero for the climate catastrophe. Its glorious Great Barrier Reef is dying, its world-heritage rain forests are burning, its giant kelp forests have largely vanished, numerous towns have run out of water or are about to, and now the vast continent is burning on a scale never before seen.

The images of the fires are a cross between "Mad Max" and "On the Beach": thousands driven onto beaches in a dull orange haze, crowded tableaux of people and animals almost medieval in their strange muteness— half-Bruegel, half-Bosch, ringed by fire, survivors' faces hidden behind masks and swimming goggles. Day turns to night as smoke extinguishes all light in the horrifying minutes before the red glow announces the imminence of the inferno. Flames leaping 200 feet into the air. Fire tornadoes. Terrified children at the helm of dinghies, piloting away from the flames, refugees in their own country.

All this, and peak fire season is only just beginning.

As I write, a state of emergency has been declared in New South Wales and a state of disaster in Victoria, mass evacuations are taking place, a humanitarian catastrophe is feared, and towns up and down the east coast are surrounded by fires, all transport and most communication links cut, their fate unknown.

Richard Flanagan, "Australia is Committing Climate Suicide"[1]

🌳 🌳 🌳

Okay, Australia is the driest inhabited continent on the planet with a history of severe bushfires, so Australians have learned to live with fire each summer. But the ferocity and extent of the 2019/20 fires were truly something else, on a scale not seen in modern times.

On the back of several years of savage drought, with temperatures often in the mid-40s and blowtorch strong winds out of the central deserts, the scene was set and a bad fire season was predicted—but not the deadly catastrophe that devastated much of eastern Australia. Many Australians were traumatised and stunned and all were appalled by the extent and ferocity of the bushfires across several states.

In his poem "The Bush Fire", Henry Lawson wrote about Flash Jim, Boozing Bill and Constable Dunn rescuing: "Swearing Pat, with his grey beard singed, and his language of lurid hue, and his tough old wife, and his half-baked kids". He also wrote:

And better the rattle of rifles near, or the thunder on deck at sea,
Than the sound—most hellish of all to hear—of a fire where it should not be.

I very much doubt that when putting pen to paper in 1905 Lawson could have conceived of the extent and destruction wrought by the 2019/20 holocaust. Those bushfires set new benchmarks. They started earlier, lasted longer, and destroyed a greater area than any on record. They also killed more than 30, and incinerated billions of dollars' worth of productive rural assets, livestock, forests and infrastructure. The impact on human welfare and economic capacity is simply too large to comprehend.

Amazing when you think about it, that a country spending billions and billions of dollars on joint strike fighter aircraft and submarines had to go cap in hand—beg and borrow firefighting aircraft from overseas. And embarrassing that three highly experienced American firefighters—captain Ian MacBeth,

Poet Henry Lawson ... "fire where it should not be".

Hercules crash site ... three American water bombing air crew—Ian MacBeth, Paul Hudson and Rick DeMorgan—have been added to the list of fallen in an out of control bushfire.

first officer Paul Hudson and flight engineer Rick DeMorgan—have been added to the honour list of the fallen after their C130 Hercules water bomber crashed while fighting an out of control blaze in the Snowy Monaro region. The aircraft was one of four large water bombing tankers deployed by Canadian firm Coulson Aviation, along with a number of helicopters, to help fight Australia's bushfires.

Long and expensive firefighting was part of the tragedy, involving firefighters from Canada, the USA, New Zealand and elsewhere, and squadrons of planes and helicopters constantly water bombing fire fronts in dangerous conditions. Crashes were on the cards and fatalities were inevitable. The military were also called into action to support the thousands of fire fighters on the ground and to rescue threatened communities off beaches—like a scene from the evacuation from Dunkirk.

Believe it or not, in January 2020 the Australian Government at long last came out unequivocally acknowledging that climate change was a root cause of the bushfire calamity. This sets the context for this book. Australia is now directly in the crosshairs of climate change. The unequivocal message from the devastating fire season is the need to act decisively. Perhaps it is already too late as climate change closes in on planet Earth and humanity.

Volunteer Brett Robin, a fifth-generation logging contractor from Gippsland, Victoria, breaks from fire-fighting duties to comfort a baby koala.

The bushfires of 2019/20 are expected to help drive global atmospheric levels of carbon dioxide to one of the greatest annual increases on record. According to calculations released by Britain's Met Office,[2] average carbon dioxide levels reached about 411.5 parts per million in the atmosphere in 2019, owing to the steady climb propelled by human activities. The bushfires are estimated to have produced at least 350 million tonnes of carbon dioxide, about two thirds of Australia's annual emissions based on burning of 5 million hectares.

The global average atmospheric carbon dioxide in 2019 was 409.8 parts per million. Carbon dioxide levels in 2021 were higher than at any point in at least the past 800,000 years.

According to the Scripps Institution of Oceanography and the National Oceanic and Atmospheric Administration, the amount of carbon dioxide in the air in May 2020 hit an average of slightly greater than 417 parts per million. This is the highest monthly average value ever recorded.[3]

The continuing rise in carbon dioxide concentrations may sound surprising in light of recent findings that the corona virus pandemic, and the associated lockdowns led to a sharp drop in global greenhouse gas emissions. Government measures during the pandemic drastically altered patterns of energy demand. Many borders were closed and people were confined to their homes, which reduced transport and changed consumption patterns.

There was an appreciable decrease in carbon dioxide emissions. By early April 2020 emissions had decreased 17 per cent compared with mean 2019 levels. At their lowest, emissions in individual countries had decreased by 26 per cent on average. Alarmingly, even with this sharp reduction, atmosphere concentrations of carbon still increased. However, the longer term impact of 2020 on atmospheric carbon levels depends on the duration of the restraint. Government actions and economic incentives after the crisis will likely influence the global carbon dioxide emissions path for decades ahead.[4] Carbon dioxide emissions reductions of the order of 20 to 30 per cent would need to be sustained for twelve months or more in order for atmospheric carbon dioxide to decline in a detectable way.

The total amount of carbon dioxide that finishes up in the atmosphere is driven not only by human-induced emission levels, but also through processes on the land surface, especially in forests and the oceans, that fluctuate on a regular basis.

In my last book *By the Light of the Sun: Trees, Wood, Photosynthesis and Climate Change*, I was delighted that Dr René Castro Salazar, Assistant Director-General, Climate, Biodiversity, Land and Water at the United Nations Food and Agriculture Organization agreed to write the foreword. This time around I approached others I thought might be appropriate experts to repeat the process, without any success. Seems people are happy to write a foreword if the book is upbeat and optimistic, but less so when it is blunt and somewhat bleak—even though it's arguable that this book is realistic. Most folk, it would seem, would rather not be associated with what they could reasonably consider a doomsday book.

One brave soul stepped forward, so I am extremely grateful to Dr John Hewson for his contribution to this book. Thank you John.

To put it mildly Dr Hewson's credentials are impressive. He has been Federal leader of the Liberal Party, and had careers in politics, academia, bureaucracy, business and the media. He is a professor in the Crawford School of Public Policy at the Australian National University, and adjunct professor at Curtin, Canberra and Griffith Universities and the University of Technology Sydney, having been Professor and Head of the School of Economics at the University of New South Wales, and Professor of Management and Dean of the Macquarie Graduate School of Management. He was a founder of Macquarie Bank, chairman of ABN Amro Australia, and chair or director of a host of companies, with current positions in insurance, renewable energy, funds management and investment banking.

He is chair of the Business Council for Sustainable Development Australia and of BioEnergy Australia, and a patron of the Smart Energy Council and the Ocean Nourishment Foundation. Dr Hewson's work on climate change and sustainability ranges from the '93 *Fightback* policy promising a reduction in emissions, through to his role chairing the National Business Leader's Forum on Sustainable Development, and the Asset Owners' Disclosure Project. Recognising opportunities in response to the climate change challenge, he has been involved in the start up of businesses in garbage recycling, energy efficient lightbulbs, bio-diesel plants, green data centres, converting sugar cane into electricity and ethanol, ultra-pure graphite for batteries and heat storage, coal refining and solar power, amongst others.

This book explores climate change issues, including the question of what might fill the ecological vacancy once occupied by humanity. Most certainly, as has happened before with other species, something will do just that if *Homo sapiens* cannot sustain themselves. We know trees have inhabited the Earth since long before humanity and will most likely continue to thrive, adapt and expand with the demise of humans. When humans move out will trees move in?

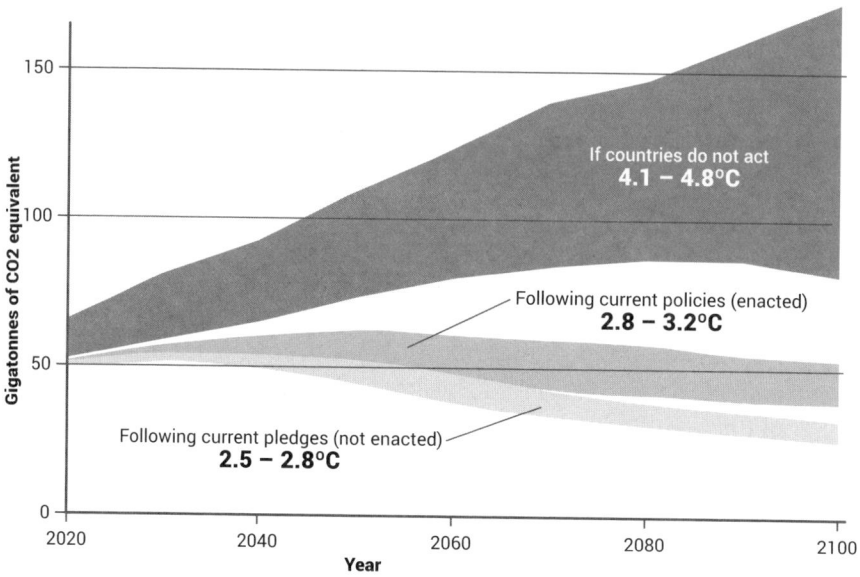

Atmospheric carbon emissions 2020 to 2100. Emissions (in Gigatonnes of carbon dioxide equivalent) and expected warming by 2100

SETTING THE
SCENE

Humanity's tenure on the planet coming to an end

In geological time terms humans have been on planet Earth less than five minutes. Very primitive life appeared at least 3.8 billion years ago, not long after the oceans formed, and after the formation of the Earth say 4.4 billion years ago. Primitive life gradually increased in complexity over the years. *Homo sapiens* appeared perhaps as recently as 100,000 years ago.

Early, simple land plants made an entry about 500 million years ago and primitive trees appeared on the scene around 370 million years ago. So, trees have been around a good while and show no signs of going anywhere. On the other hand, if we read the science and physical signs, humanity's short tenure on the planet looks as if it may be coming to an end.

🌳 🌳 🌳

Before the Industrial Revolution human populations were relatively stable and population growth was pedestrian. Since then things have gone awry. Advances in medical science, sanitation and lifestyle factors have contributed to a surge in population, greater affluence and longer lives. Increased consumption, industrial activity and improved health have all put pressure on the natural world and its resources.

LEFT: Down out of the trees … perhaps 80,000 years ago. RIGHT: Burning coal. Since the Industrial Revolution energy use has been based on the burning of fossil fuel.

Since the Industrial Revolution, manufacturing, energy use, and transport have all continued to be pretty much founded on the burning of fossil fuel with the resultant pumping of carbon residues into the atmosphere. This has led to human-induced warming of the biosphere, commonly called global warming or climate change. Unwillingness or inability to do anything meaningful about it—or just bloody mindedness—is clearly evident. This largely because tackling climate change will of necessity mean reducing lifestyle conditions—less consumption, reduced energy use and so on—all politically unpopular.

The impacts of climate change are now readily apparent—climbing temperatures, rising sea levels, increased storm severity and so on. Increasing bushfire frequency and ferocity have the potential to result in international anarchy and a breakdown of law and order. Yet we seem unwilling or unable to get serious about these threats.

The consequences of inaction are already making themselves felt in the form of more extreme weather events and associated disasters, including hurricanes, drought, floods and fires. Ice caps are melting—in Greenland alone, 197 billion tonnes of ice were reported to have melted just in July 2019.[1] The Greenland tundra faced its worst known wildfire in 2017. Permafrost in the Arctic is thawing 70 years ahead of projections. Antarctica is melting three times as fast as a decade ago. Sea levels rising quicker than expected pose risks to the world's coasts, where more than two-thirds of the largest and most economically important cities are located.

While the oceans are rising, they are also being poisoned. Oceans absorb more than a quarter of the carbon dioxide in the atmosphere and generate about half our oxygen. Absorbing more and more carbon dioxide acidifies the oceans and threatens the future of marine life. If current trends continue, we are looking at global heating perhaps just a bit short of 4 degrees Celsius by the end of the century. The impact of that on all life on the planet—including ours—would be catastrophic. Comfortable rules-based civilisation would be likely to break down. This may signal the beginning of the end of human existence here on planet Earth.

Although this is predictable—and certainly unfortunate for humanity, the planet will be agnostic to it. The Earth has seen it all before with major orders of life evolving then disappearing as environmental conditions change, suddenly as in the case of dinosaurs, or more gradually as ice has advance in the coming of the ice age, then retreated. The difference this time is that climate change is being induced by life—you and me—and its impact on us will likely be fatal.

Bushfire in 2020. Emerging impacts of climate change include increasing bushfire frequency and ferocity.

The end of humanity should not really come as a surprise—climate change notwithstanding. The reality of life on Earth is that species of plants and animals continue to evolve, flourish and disappear, for a variety of reasons. Changes in physical and climate conditions have initially favoured, then disadvantaged, particular individual species. Life emerged, emerged, evolved and adapted long before humanity put in an appearance, and will most certainly continue to do so after humans have gone the way of the dinosaurs. We will get to that.

So, what will fill the ecological space currently occupied by humanity? Something, certainly will.

Are trees the natural inheritors of the Earth, the silent sentinels that stand guard over enduring life on our planet? They were present long before the early ancestors of humanity clambered down from the trees and walked on Earth, and they will most likely continue to thrive, adapt and expand when humans have exited the stage. Trees are bigger, live longer and are more resilient than members of the animal community. This book examines the proposition that trees are much more important than humanity to planet Earth, and much more able to withstand climate change extremes, whether natural or human induced.

Increased atmospheric carbon dioxide levels arising from climate change may even be good for some trees and other plant growth, at least in the short term. There is solid evidence in the pages ahead that trees will quickly be able to reoccupy areas from which they have been evicted

LEFT: As in the case of dinosaurs, the planet will be agnostic to species demise. RIGHT: Prehistoric man. Trees were present long before the ancestors of human beings clambered down from the branches and walked on Earth.

Hotel Polissya, one of tallest buildings in abandoned apocalyptic Pripyat, hit by the nuclear disaster of Chernobyl power plant.

by humanity, including quite rapidly invading areas with a hard human footprint, such as big cities.

From the case studies examined later—the Chernobyl nuclear power plant area and the Demilitarized Zone separating the two Koreas—we learn that many forest-dwelling animals precariously perched on the edge of extinction are likely to reassert themselves when the human pressure on their habitat is removed. So good news for forests and forest animals. They will be the great beneficiaries of our departure.

Tree communities have long been here and are likely to be the next masters of the Earth if humanity's fleeting appearance ends in self destruction. They are also unlikely to do as humans are doing and damage the biosphere to such an extent that they cause their own departure.

ENCOUNTERS FROM OUT THERE IN SPACE

Death of the dinosaurs— might it happen again?

Space is a big place with lots of things whizzing around at super high speed. Not surprisingly, from time to time they bump into each other with powerful and spectacular results. But more often than not objects travelling through the vastness of space speed past each other with no dramas. At times, in the blackness of space, asteroids, comets and other objects miss each other in close encounters. For instance on 25 July 2019, an asteroid named 2019 OK, travelling at almost 25 kilometres a second, came unusually close to planet Earth. It passed by about 70,000 kilometres away—a fair distance you might think, but closer than the Moon. It was one of the closest known asteroid fly-bys since we began tracking objects moving through space in our neighbourhood of the Milky Way.

An asteroid like 2019 OK would have had a spectacular impact, and caused horrific damage had we been on the receiving end of a direct hit. Scientists reckon that 2019 OK would have struck Earth with over 30 times the energy of the atomic blast at Hiroshima, even though as flying objects in space go it wasn't all that big.

A near miss by 70,000 kilometres is very close compared with the vastness of space, but stay calm. It's still a fair way off. The Earth itself is about 13,000 kilometres across and only about half a percent of asteroids that come this close or closer will actually bump into us.

Fortunately, most asteroids are small and burn up in the atmosphere as shooting stars. A few are large enough to retain some solid mass all the way down to the Earth's surface, and once in a few tens of millions of years, a very large one collides with us with catastrophic results. Because of their high velocity relative to Earth, such massive objects release an unimaginably quantity of energy when they impact.

Should an asteroid like 2019 OK hit us in a populated locality, the casualties would likely to be greater than those from most natural disasters. Yet so much of our planet's surface is ocean, desert, or remote wilderness that the odds of even a small asteroid striking one of our major cities are fairly slim.

While impacts from asteroids the size of 2019 OK are very uncommon, they are not unheard of. An asteroid about that size hit Earth a bit over a century ago, luckily in the remoteness of Siberia. Its impact was monumental as it destroyed about 2,000 square kilometres of forest. Estimates suggest that an asteroid like that will enter Earth's atmosphere about once every 300 years.

What could we do if we saw a really big piece of rock hurtling in our direction from out there in space? Panic, I guess! Scientists have thought about it, and say it wouldn't make a lot of sense to launch the world's entire nuclear arsenal at it in the hope of blowing it up—even though some world leaders might be inclined to do so. You'd need an astounding amount of force, and you'd need to apply it in just the right spot to explode the asteroid into fragments. Knocking the asteroid off course looks a little more realistic. But our best defence here on Earth is the sheer unlikeliness

LEFT: The asteroid 2019 OK passed Earth by about 70,000 kilometres. RIGHT: Fortunately, most objects entering from space are small and burn up in the atmosphere as shooting stars.

that a very large asteroid will hit us in the first place. As we will see, humanity is far more likely to become extinct through our own actions than through a very large rock arriving from space, and as individuals, we're in much greater danger from walking across the street than from rocks falling out of the sky.

Let's change tack and consider life here on our home in the cosmos. Life and evolution are slow but predictable, incremental processes. However, there have been instances over the hundreds of millions of years where evolution has accelerated or taken an unexpected deviation. Yes, emerging and evolving life here is, and has been, predictable and pretty much the basis of the slow and steady evolutionary process that saw humanity evolve from primates. While we now applaud the genius of Charles Darwin who first provided the science to support this understanding and discount religion-based creationist doctrines, we also know about the predictable yet fragile nature of life here on Earth.

TOP: Massive asteroid impact that killed the dinosaurs. ABOVE: Charles Darwin, who controversially at the time argued that humans had descended from animals.

For example, the echo-ranging bat is the result of an inching series of minor improvements, each adding cumulatively to its predecessors as the evolutionary trend is propelled forward. The theory of evolution explains complex biological organisms that project sophisticated design and development.

Given that we humans are mammals, we should be interested in what happened about 65 million years ago when animal life crossed the

During the Age of Dinosaurs mammals that existed were small and insignificant—mice, rats and shrew-like fur-covered creatures.

threshold which separates the Age of Mammals from the much longer Age of Dinosaurs. Until that time mammals were generally small and insignificant—mice, rats and shrew-like fur-covered creatures, their evolutionary potential held down under the reptilian supremacy of the dinosaurs for more than 100 million years. Suddenly that pressure was released and, in a very short time geologically speaking, the descendants of those somewhat innocuous mammals expanded to fill the ecological spaces once occupied by millions of dinosaurs.

A few theories as you might expect have been proposed to account for the very sudden disappearance of dinosaurs. It was thought there was extensive volcanic activity in the area now called India, spewing out lava flows covering well over a million square kilometres which would have had a radical effect on the climate and landscape. However, the final death blow that wiped out the dinosaurs was something much more sudden and drastic.

It is now known that a massive asteroid or perhaps a comet hit the Earth. The sort of impact we are talking about would have disintegrated the incoming projectile, and scattered its remains as dust throughout the atmosphere, which would eventually have rained down all over the Earth's surface. The footprint, more than 150 kilometres wide and 50 kilometres deep—is a titanic impact crater, Chicxulub, at the tip of the Yucatan peninsula in Mexico.

The asteroid impact that created the titanic impact crater, Chicxulub, at the tip of the Yucatan peninsula in Mexico would probably have deafened every living creature not incinerated by the blast, suffocated by the wind-shock, drowned by the 150m tsunami that raced around the literally boiling sea, or pulverised by massive earthquake, and that was just the immediate calamity.

The noise of the impact—thundering round the planet as it would have—probably deafened every living creature not incinerated by the blast, suffocated by the wind-shock, drowned by the 150-metre tsunami that raced around the literally boiling sea, or pulverised by an extraordinary massive earthquake, and that was just the immediate calamity. There would have been the aftermath—the global forest fires, the smoke and ash which blotted out the Sun for perhaps a couple of years, killing most of the plants and destroying the world's established food chains.

The dinosaurs, most too big to hide or adapt, perished—and not just the dinosaurs, but about half of all other species too, especially marine species. The wonder is that any life at all survived this cataclysm.

Early large mammal … the age of the mammals had arrived.

In evolutionary terms it was time to start afresh. This time, an agile minor player, a vertebrate order called *Mammalia*, had its chance to make a move. The age of the mammals had arrived.

IT'S HEATING UP
HERE ON
PLANET EARTH

The pall of smoke obscuring Sydney is a grim reminder that the impact of climate change is pervasive and all too real. For burning of fossil fuels, deforestation and increased agricultural production did not cause the bushfires or the protracted drought affecting Australia, but they have caused global warming which has, in turn, exacerbated the scale and intensity of these calamities.

Even people who have ignored the issue are getting an inkling that climate change is not just a political debate, but something that will affect our daily lives. The hazy skies are a warning that it will change the great Australian summer.[1]

As politics and economics have increasingly come to dominate our decisions and actions, we humans have lost our sense of place in the world, and our reverence for nature. We think we really are big time, but in fact, planet Earth doesn't care a toss about us—or any other form of life for that matter. The inner workings of the Earth remain unaffected by the life bubbling away on its surface, and if we self destruct, the planet will carry on without missing a beat.

It is gravity that is a fixed condition of a planet, and the workings of the Earth which of course life cannot influence. It's the Earth's distance from the Sun—and other realities like its speed of rotation in the solar

Workings of the Earth remain unaffected by the life bubbling away on its surface.

system—which determine day length. On a planet like ours with a near circular orbit, it's the tilt of the Earth on its axis that mainly determines the seasons.

By contrast, on a planet with a far-from-circular orbit, like Pluto, the dramatically changing distances from the central star are a much more significant determinant of seasonality. Anyway, Pluto is much colder than freezing.[2] Moons, in our case just the one, also exert an influence on conditions on the surface of planets, also corresponding to their distance, mass and lunar orbits—as our Moon does with ocean tides. All these factors are permanently fixed and not influenced by the biological happenings on the surface.

Yes, there will be an end game for our Blue Planet in around five billion years from now, when the Sun will expand to become a red giant, absorbing all the inner planets into its fierce and fiery belly. The consequences will reach across the solar system; perhaps the ice will thaw on Saturn's moon Titan, where the temperature is currently minus 150 degrees, and some interesting creatures may crawl out of its methane lakes.

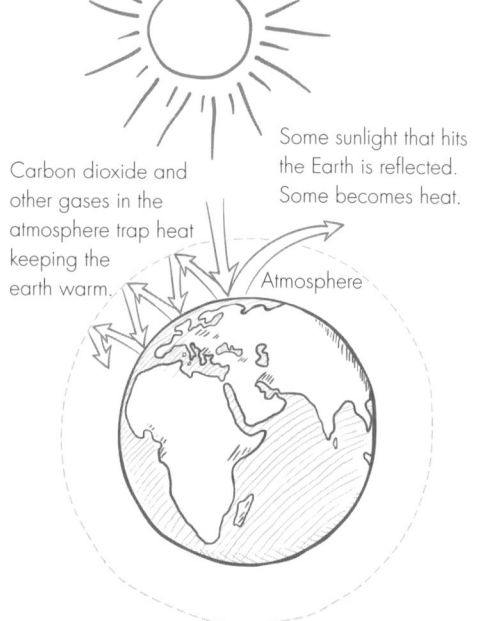

Carbon dioxide and other gases in the atmosphere trap heat keeping the earth warm.

Some sunlight that hits the Earth is reflected. Some becomes heat.

Atmosphere

ABOVE: Planet Earth … our rock in the solar system. RIGHT: The greenhouse effect. Human activities have increased the levels of greenhouse gases, trapping heat in the atmosphere and causing global temperatures to rise.

Yet, not every condition on Earth is completely beyond the influence of life forms. The atmosphere, especially its oxygen and carbon components, is a result of the influence of the evolution of life in all its various forms over vast geological periods. Life has influenced, and continues to influence the weather, and even major climatic episodes, such as ice ages and droughts.

Without the atmosphere, the Sun's heat would just rebound off the Earth's surface back into space. It would be some 30 degrees Celsius colder here on Earth and everything would be frozen solid. The atmosphere acts like the glass sides of a greenhouse. Hence the warmth it gives us is referred to as the *Greenhouse Effect*. The so-called *greenhouse gases* in the atmosphere trap heat and are responsible for this effect.

Most greenhouse gases occur naturally, including carbon dioxide, methane and water vapour. More potent greenhouse gas molecules like chlorofluorocarbons, "CFCs", are manufactured chemical compounds. CFCs contain chlorine, fluorine and carbon. Because they destroy the ozone layer, CFCs have been banned, but they were used widely until recently as refrigerants, aerosol propellants and solvents.

After-effects of major volcanic eruptions and other natural phenomena cause fluctuations to global temperatures.

27

Bethlehem Steel Works in May 1881. Watercolour by Joseph Pennell. Since the Industrial Revolution human society has been producing greenhouse gases in ever increasing amounts.

Greenhouse gases allow sunlight to penetrate the atmosphere, but acting like a blanket, they reflect infra-red heat wavelengths back onto Earth's surface. The continued addition of greenhouse gases to the atmosphere thickens this blanket effect with potentially catastrophic ecological consequences.

Everything is connected in nature, so as heat-trapping gases are added to the atmospheric greenhouse blanket, polar ice sheets begin to melt, ocean waters warm, and terrestrial ecosystems begin to change as animals and plants move in order to remain within their temperature comfort zone.

The scientific evidence is now compellingly crystal clear. Past ice ages have alternated with warmer periods, and the average temperatures on Earth have varied between about 9 degrees and 22 degrees. The current average temperature is 15 degrees. Slight variations in the Earth's orbit, changes in the Sun's activity and the after effects of major volcanic eruptions can cause fluctuations.[3]

The world we know is a result of the climate remaining stable for the past ten thousand years or so, with only small natural changes of less than one degree Celsius in any country. These settled conditions have enabled us to exist in stable communities and lifestyles. But now that natural factors are no longer the main cause of global warming, temperatures are changing, and always getting hotter overall. In addition, the concentrations of carbon dioxide in the atmosphere today are the highest they have been for perhaps 650,000 years.[4]

Since the Industrial Revolution of the 18th century, human society has been producing greenhouse gases in ever-increasing amounts. Human induced climate change is well and truly underway. Since 1850, the average global temperature has already increased by one degree Celsius.

The rate of temperature increase has gone from 0.1 degree per decade over the last hundred years to 0.2 degrees in the last decade. It is predicted that the average global temperature is most likely to continue to increase

TOP: Tongue of the Breiðamerkurjökull glacier as it retreats leaving floating icebergs, Vatnajökull National Park. A few degrees change makes a big difference to the climate. ABOVE: By removing natural forests over the last three centuries between 200 and 250 billion tonnes of carbon have been released into the atmosphere, between 22% and 43% of all carbon released in that period.

to between 1.8 degrees and 4.0 degrees over the course of this century, and that it could rise by as much as 5 degrees.[5]

Temperature rises of this size may not seem huge until you remember that during the last ice age, which ended 11,500 years ago, the average global temperature was just 5 degrees cooler than today's, yet polar ice covered much of the Northern Hemisphere.[6] A few degrees change in average temperatures makes a heap of difference to the climate.

The evolution of life over geological time demonstrates a fair amount of flexibility, and given sufficient time evolution can adapt life to changing

San Jacinto, Manabi, Ecuador. High tides, combined with rising sea levels damage the coast. Global warming and rising sea levels will decide the fate of hundreds of thousands of species, and billions of people.

climatic conditions. So, it is the speed of warming, rather than the direction or overall scale of change, that is important. Climate scientists argue that warming rates above 0.1 degrees per decade will rapidly increase the risk of significant ecosystem damage. Similarly, rates of sea level rise above two centimetres per decade would be potentially very dangerous, especially if you live close to the ocean.[7]

The reality is that the carbon concentration in the atmosphere continues to grow, at a faster rate than before. Two hundred years ago the atmospheric concentration of carbon as carbon dioxide was around 2.8 parts per ten thousand. Today it's more than 4.0 parts by some measures—a level not seen for at least three million years. Even if we ceased burning fossil fuels today, it would take several centuries for the oceans, the Earth's crust and biological life to re-absorb the excess carbon.[8] But that will not happen. Instead, according to some predictions, we may be on track if we do nothing, to increase the concentration to at least 7.0 parts per ten thousand by the end of the century.

For the first two hundred years following the commencement of the Industrial Revolution, emissions increased at an average of 2 per cent per year. Since 2000 they have increased by an average of 3.4 per cent per year, a rate that is driving atmospheric carbon dioxide levels beyond pessimistic predictions. By removing forests across the planet over the last

Melting glacier in Antarctica. Evidence reveals the permafrost thawing and glaciers retreating.

three centuries we have released between 200 and 250 billion tonnes of carbon in the atmosphere making up between 22 and 43 per cent of all the carbon released during that period.[9]

There is now no doubt that climate change will increase the intensity and frequency of extreme weather events, such as storms, floods, droughts and heat waves. Water is already scarce in many regions. Almost one fifth of the world's people do not have access to clean drinking water. If global temperatures increase by 2.5 degrees above pre-industrial levels (that is, around 1.7 degrees above present levels), an additional three billion people might suffer from water scarcity. Rising sea levels will almost certainly lead to regional conflicts, famines, and the mass movement of climate change refugees, as food, water and energy resources become scarce.

We should be under no illusions as to what is at stake. The Earth's average temperature is around 15 degrees, and whether we allow it to rise by a single degree, or 3 degrees, or more will decide the fate of hundreds of thousands of species, and billions of people.

Our lack of action indicates that we don't appreciate the seriousness—the near and present danger—of the climate change threat to continuing life. Let's recap for a moment. There is no longer any doubt that the climate is changing due to human activities, especially the burning of fossil fuels. According to the United Nations Intergovernmental Panel on Climate Change or IPCC,[10] human-related activities contribute six billion tonnes

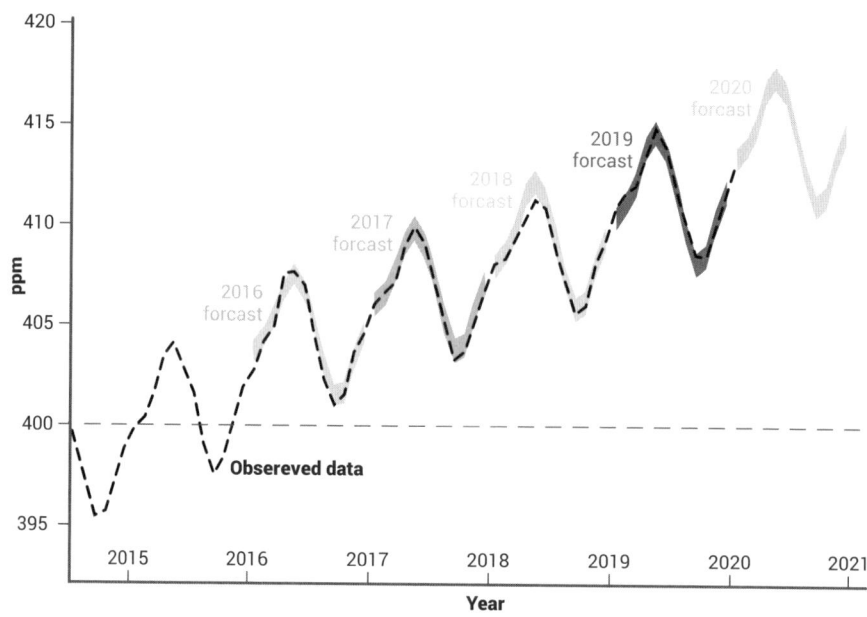

Atmospheric carbon dioxide concentrations

of carbon emissions annually, growing by half a per cent a year. The most optimistic estimates indicate that the concentration of carbon dioxide in the atmosphere will double by 2100.

Industrial methods of "carbon capture" are still being researched and are yet to be operationally tested. There is no doubt that plants remain the most effective mechanism in existence for capturing carbon. Each year plants deal with eight per cent of the atmospheric carbon dioxide. Trees and wood products are part of this conjuring act storing this captured carbon longer term.

A curious reality of global warming is that as Earth warms, places where cold inhibited plant growth become steadily more hospitable. The new vegetation exerts its own effects on the climate. According to a team led by Trevor Keenan of the Lawrence Berkeley National Laboratory in California,[11] this plant growth caused by climate change may also be helping to slow climate change—at least temporarily.

According to Keenan's team, between 1959 and 1989, the rate at which carbon dioxide levels were increasing in the atmosphere rose from 0.75 parts per million per year to 1.86 parts per million per year. Since 2002, this trajectory has stalled. In other words, although more carbon dioxide than ever is being pumped into the atmosphere, less than might be predicted of this excess is staying in the atmosphere.

Towards the end of the 20th century around 50 per cent of the carbon dioxide emitted each year was removed from the atmosphere and stored in carbon sinks. Now, according to research led by Keenan, the rate of removal has grown closer to 60 per cent. Carbon sinks seem to have become more effective, but the precise details are still unclear.

Using a mix of ground and atmospheric observations, satellite measurements and computer modelling, the Keenan team concluded that faster-growing plants are the chief reason for this increased storage of carbon in sinks. The basic proposition is that as carbon dioxide concentrations rise, photosynthesis speeds up. Studies conducted in greenhouses have found that plants can photosynthesise up to 40 per cent faster when concentrations of carbon dioxide are between 475 and 600 parts per million than they can in natural conditions.

However, Keenan's team sounds a warning—more vigorous photosynthesis is only temporarily slowing climate change and will not last. There is more to growing plants than carbon dioxide. Climate change probably means areas of high rainfall are becoming wetter, while drier areas are becoming drier. Changing rainfall patterns could make some places less friendly to plants, and although plants benefit in the short term from extra carbon dioxide they suffer when temperatures get too high.

There will be other more complicated effects. For instance, much of the new plant growth has occurred in colder regions of the Earth and while ice and snow reflect sunlight, vegetation soaks it up, so more greenery in the north of the Northern Hemisphere will eventually lead to yet more warming. Elsewhere, higher temperatures could damage tropical forests. According to one estimate, for every degree of warming, tropical forests may release greenhouse gases equivalent to five years' worth of human emissions.

As carbon dioxide concentrations rise, photosynthesis speeds up. Studies in greenhouses have found that plants photosynthesise up to 40% faster when concentrations of carbon dioxide are between 475 and 600 parts per million.

"Global greening, then, offers only a little breathing space," wrote Chris Mooney, summarising the Keenan research in *The Washington Post*,[12] "kicking the fossil-fuel habit remains the only option."

We know that the first effects of adverse climate change have already occurred, and point the way to much more widespread and destructive changes in the future. Detrimental climate change impacts include the reality that the North Pole ice cap is melting and research tells us that its surface area has diminished by 20 per cent between 1950 and 2000.[13] Evidence also points to global examples of snow cover diminishing, glaciers retreating and permafrost thawing.

Sea levels rose by about 15 centimetres in the 20th century. There has also been a significant increase in the frequency and severity of natural disasters, such as hurricanes, droughts and floods. Globally the increase in flood damage over recent decades has been a real eye opener. In the 1960s about seven million people were affected by flooding each year. Today that figure stands at 150 million.[14]

The obvious message is that unless climate change is brought under control—and that is looking less and less likely—it will lead to serious catastrophic events, like continuing rising sea levels, food and water shortages and civil unrest. While climate change will affect all countries, developing nations are the most vulnerable. To a large degree these countries depend on climate-sensitive activities, especially agriculture, and have only limited capacity to adapt to the consequences of future adverse weather patterns.

DROUGHT, FLOODS AND BUSHFIRES

Understanding the Australian condition

George Goyder described drought as "insufficient water to meet needs".[1] He said native plants that grow abundantly in a region a farmer describes as drought-prone have sufficient water for their needs. It might reasonably be argued that Australia does not suffer from drought, but from attempts to grow inappropriate plants. Charles Darwin recognised this. He wrote in his diary, during his eighteen day stopover in Sydney in January 1836:

> *Nowhere is there an appearance of verdure and fertility, but rather that of arid sterility. I cannot imagine a more complete contrast in every respect than the forest of Valdivia or Chiloe, with the woods of Australia. Although this is such a flourishing country, the appearance of infertility is to a certain extent the truth. The soil without doubt is good, but there is so great a deficiency in rain and running water, that it cannot produce much. The pasture everywhere is so thin that already settlers have pushed far into the interior; moreover, very far inland the country appears to become less profitable. I have before said, agriculture can never succeed on a very extended scale.*[2]

The truthfulness of Darwin's observation is reflected in the experience of farmers that Australia is a drought-prone country. Studies claimed that

A Cuban stamp depicts the expedition route of *HMS Beagle*, carrying Charles Darwin.

perhaps two thirds of Australia is uninhabitable due to lack of water, a conclusion also backed by geographer Griffith Taylor in his book, *Australia in its Physiographic and Economic Aspects.*[3]

As far as drought experienced by farmers is concerned, Australian records do not seem to show much long-term change—although the first decade of this century does look to be rather more under the influence of drought than any earlier decade. Droughts are compared across regions and times using the Palmer Drought Severity Index,[4] a standardised measure ranging from about minus 10 (dry) to plus 10 (wet), of surface moisture conditions calculated from readily available temperature and rainfall data. The diagram opposite shows this index for the Riverina district in south-eastern Australia, while Australia's major droughts are also shown.

A cow succumbs to drought in the Northern Territory.

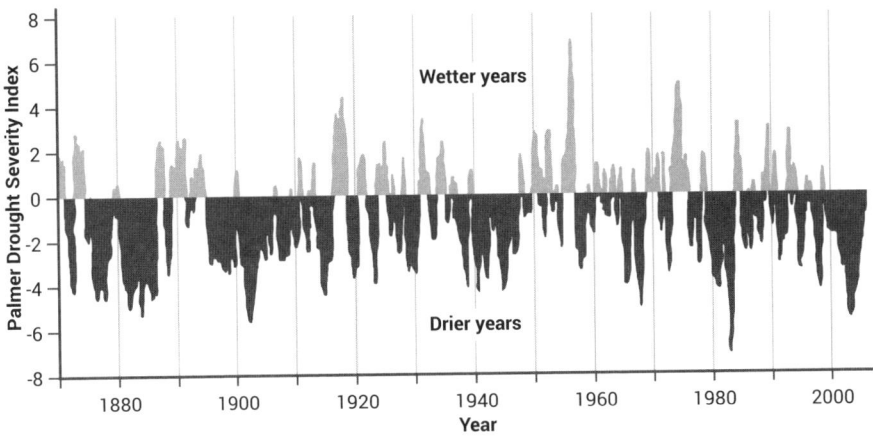

Drought years in the Riverina up to 2005. Drier years are shown below the line with higher negative values indicating more severe droughts.

Research by Caroline Ummenhofer, Associate Scientist, and colleagues at the Physical Oceanography Department at Woods Hole Oceanographic Institute, Woods Hole, USA, analysed the occurrence of droughts in South East Australia and linked them to changes in the Indian Ocean. The conclusion was:

> *When taken in the context of other historic droughts over the past 120 years, the Big Dry (the drought from 1998) is still exceptional in its severity. Furthermore, the severity of the Big Dry has been exacerbated by recent warmer air temperatures over the past few decades. Warmer air temperatures lead to increased evaporation, which further reduces soil moisture and worsens the drought.*[5]

While this work does not explicitly focus on the link between changes (in the Indian Ocean) and regional and global warming, it does send a stark message: "in a warmer world, the severity of droughts would likely become far worse."[6] This is consistent with experience in the 2019/20 drought and bushfires as outlined in Chapter 1.

Scientists from the National Center for Atmospheric Research in Boulder, Colorado, United States of America published their analysis of worldwide soil moisture changes from 1870 which showed that since 1950 there has been a tendency for more severe drought, most notably across southern Africa, and for heavier rainfalls in parts of North and South America.

As is hardly surprising, these results mirror the worldwide rainfall compilations and emphasise that climate change is a reality.[7]

LEARNING TO LIVE IN A GREENHOUSE

Consequences of imminent changes to weather patterns

After last chapter's exploration of the effects of greenhouse gases, we need to get down a bit further into the role of human activity in producing carbon dioxide emissions, and some of the dangerous consequences. Carbon dioxide emissions as a result of human activity amount to about 25 billion tonnes per year over the past decade, around four tonnes each year per person.[1] Climate scientists point to the absolute necessity of halving this rate by the middle of this century if already apparent climate change trends are to be kept within even somewhat tolerable limits. In my view we are simply not going to achieve such a goal.

🌳🌳🌳

At least half of all present carbon dioxide emissions can fairly be attributed to modern industrial countries. At 12.6 tonnes per person, carbon dioxide emissions in these countries are five to six times higher than in developing countries, where emissions average 2.3 tonnes per person. Amongst developing countries there is a large variation from less than

Industrial pollution responsible for significant greenhouse emissions. Halving this rate by the middle of this century is an absolute necessity.

a tonne per person in the poorer countries to 4.5 tonnes in those with increasing personal income levels. The spectrum is also wide in modern industrialised countries, from 5.5 tonnes in Malta and Sweden to a whopping 20 tonnes in the United States of America. The level of emissions per person is a staggering 200 times greater in the United States than in some countries of central Africa.[2]

To a substantial degree all this is driven by energy consumption. Four fifths of the energy use in developed countries comes from ancient organic waste, what we commonly call fossil fuels. This is carbon that nature didn't need, so it was tucked away safely on or under the ground. Over aeons, this buried material was compressed into highly concentrated coal and petroleum. Then, in less than three centuries we humans have dug up hundreds of millions of years' worth of nature's carbon bank and burned it. It's the exhaust from this burning that has filled the atmosphere with more carbon dioxide than has beset the Earth in at least three million years.

A farmer tending her animals in Ethiopia. Emissions per person in the USA are 200 times more than in some African countries.

LEFT: Layers of coal and petrified wood in a canyon wall. Over geological time huge quantities of carbon dioxide have been stored in or below Earth's surface when dead plants are buried and become fossil fuels. RIGHT: Vehicles in Italy in a traffic jam. The exhaust from burning petrol has contributed to higher carbon volumes than the atmosphere has contained in at least three million years.

Carbon is constantly shifting in and out of our bodies, as well as between rocks, sea and soil, into the atmosphere and back again. As most carbon exchanges involve carbon dioxide, what are commonly known as *carbon sinks* are able to capture carbon dioxide in the cycle and to reduce its concentration in the atmosphere.

Huge quantities of carbon dioxide have been stored in or below Earth's surface, as dead plants are buried and become fossil fuels. Tonnes of carbon are tied up in living things and oceans, while the amount buried underground is far, far greater.

Soils represent a huge carbon reserve, around 150 billion tonnes worldwide. However, the world's intensively used crop lands have lost 30 to 75 per cent of their carbon content over the past two centuries. That is around 78 billion tonnes of carbon. When combined with the carbon lost from poor land management and from eroded soils, it is clear that a huge amount of carbon has moved from soils into the atmosphere.

Above ground carbon reserves are largest in the tropical forest belt that circles the equatorial centre of the planet. Tropical forests, covering only about five per cent of the Earth's surface, are disproportionately high in importance to the climate system. In addition to storing carbon they have a major influence on global weather patterns—and also of course on biodiversity, because an estimated two thirds of all living species reside in these forests. The significant consequences of their destruction are clear.

We now know all too well that the consequences of runaway climate change will be horrific. Rising sea levels will have an uneven impact, first pushing politically and economically weak regions, like tiny South Pacific island nations, to the limits of survival—or beyond. At least 100 million people live a metre or less above sea level. Sea levels could rise by a metre in the foreseeable future. What will happen to these folks? Where will they go? Who will offer them a new home? Will they all just perish? Do we care anyway?

We know that some greenhouse gases are long-lived, meaning they hang around in the atmosphere for decades. So even if we take strong action today, temperatures will continue to rise in the immediate future. Little action will mean that temperatures will increase even more. The price of negligible or ineffective action will obviously be much higher because of the damage and human suffering that unconstrained climate change will unquestionably cause. Even if decisive action is initiated to cut greenhouse gas emissions savagely many of the changes that are already underway simply cannot be halted.

In developing countries crops will need to be found that need less water, and can tolerate rising temperatures. For already drought-prone countries like Australia, technologies to use water more frugally and bushfire mitigation and suppression demand science input and increased resourcing. Housing and building construction will need increased emphasis on storm resistance and energy efficiency.

Coastal erosion at Skipsea, UK. Rising sea levels will impact unevenly around the planet.

LEFT: Bushfire mitigation and suppression will need scientific input and increased resourcing. RIGHT: An out of control bushfire in southern New South Wales. Hot and dry conditions over the 2019/20 summer caused the destructive fires in eastern Australia.

The incidence of extreme weather events, the melting of the Arctic ice cap, and the acidification of the oceans caused by the absorption of carbon dioxide are all increasing faster than was anticipated. The situation is now so entrenched that an Australian report[3] notes that the global average surface temperature is unlikely to drop in the first thousand years after greenhouse gas emissions are cut to near zero.

Am I scaring you? Good. Some of this is bleak for sure. However, it is well past time just to be an alarmist. Spelling out the facts will help us to confront climate change challenges squarely and turn our attention to fashioning workable solutions that may help some of us survive.

At home here in Australia warming sea temperatures are threatening the Great Barrier Reef with widespread coral bleaching. Prolonged hot temperatures have also contributed to major algae blooms in inland rivers, and hot and dry conditions over the 2019/20 summer were considered a major factor in destructive fires that were exacerbated by severe drought conditions over most of eastern Australia.

Coral reefs take years or even decades to recover after severe bleaching. Repeated bleaching can kill the coral and turn the reef into an algae-dominated ecosystem.[4] The impact of hotter conditions on

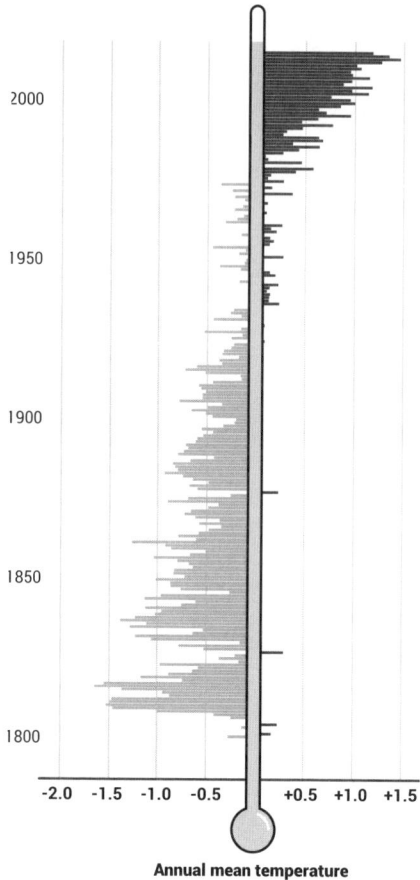

Annual mean temperature

The world is getting warmer. Annual average land temperature. The average is calculated from 1951–1980 land surface data.

high-profile ecosystems also comes at an economic cost. The Great Barrier Reef delivers billions of dollars of value-added economic activity every year. The aftermath of horrific bushfires is having a depressing impact on industries, tourism and related businesses up and down eastern Australia.

The average global land and ocean temperature in January 2019 was 0.88° above the 20th century average, tying with 2007 as the third highest since global records began in 1880. Only the years 2016 at 1.06°, and 2017 at 0.91° were warmer.[5] The temperatures on land were particularly high, with the average land surface temperature more than 2° above the 20th century average.

2019 was the second hottest year ever, capping off the hottest ever decade. Temperatures in 2019 were slightly cooler than in 2016, which was the hottest year on record at 1.2° above pre-industrial levels. Worsening extreme heat events in Australia and around the world take their toll on human health, the environment, agriculture, infrastructure and many other facets of daily life.

Arctic sea ice typically reaches its maximum extent for the year in mid to late March. The ice cover on the Arctic Ocean has a major influence on the broader global climate system. The expanding area of summer ice melt exposes more darker ocean water, which absorbs, rather than reflects incoming sunlight and warms the region even further. For this reason, a rapid decline in the extent of sea ice is of major concern for both the Arctic and the global climate system. There is more about Arctic ice melt and the worrying climate change consequences in Chapter 16.

The environmental and economic trauma already happening reinforces the urgency of calls for immediate counter-measures. Yes, there is lots of

A polar bear navigating amongst ice floats in the Arctic. The rapid decline in the extent of sea ice is of major concern for both the Arctic and the global climate system.

talk about action to reduce emissions, which sounds good politically. But I fear the action may be too little, too late. Like it or not we will need to learn to live, if we can, with the consequences. This means determining the probable effects of an altered climate and taking action to minimise the impacts on human survival beyond this century.

DID WE KILL OFF OUR COUSINS?

From Neanderthals to Rwanda

The historical record makes it plain that as long as we have existed humans have been an aggressive, colonising, inhospitable, brutalising, warlike species. That does sound a bit tough, but go back as far as you like, this behaviour began to manifest when human species started to emerge. Like our kin the chimpanzees, we murder one another over territory, resources and mates.

According to the evolutionary psychologist John Tooby[1] people everywhere are "identity crazed". He says we can't help it. "We are wired from birth to tell *Us* from *Them*. And we inevitably and sometimes unconsciously favour *Us*—especially when we feel threatened."

Of course, humans share this *Us* from *Them* trait with many other creatures large and small. However, other creatures typically don't overturn the group identifications that determine their behaviour to each other. The birds and bees kept to their tribes while in the Balkans feuding people teamed up as Yugoslavs and later split into warring Croats, Serbs and Bosnians with homicidal intentions towards one another.

The Egyptians, Romans, Vikings, Mongols, and so on, all put their expansive aspirations into action by killing and subjugating others in the name of religion, culture, economic advancement or some other cause.

This reality was replicated around the Earth as Spanish, Portuguese, Dutch, British, French, Belgian and German explorers and subsequent

LEFT: Corpses at Germany's Belsen Concentration Camp in 1945. They were among thousands that lay unburied when the camp was liberated by Allied Forces. In the Holocaust Nazi Germany systematically murdered around two thirds of Europe's Jewish population. RIGHT: A re-enactment of the Roman invasion of Britain at Old Sarum Fort, Wiltshire.

colonisers from the Old World sought to subjugate indigenous populations.

In more modern times, in the middle of the 20th century, we can point to Germany's Nazi regime that murdered six million Jews and other so called "inferior races" between 1941 and 1945. Known as the Holocaust, the World War II genocide of European Jews saw Nazi Germany and its collaborators systematically murder around two thirds of Europe's Jewish population. The murders were carried out in mass shootings and by extermination in gas chambers in concentration camps.

European Jews were targeted for extermination during the Holocaust era, but Nazi Germany also persecuted and murdered other groups, including ethnic Poles, Soviet citizens, and prisoners of war, the so-called incurably sick, political and religious dissenters, and gay men. Not counting enemies killed in warfare, the slaughter of these groups is thought to have pushed Germany's death toll to 11 million.

Over recent decades we have seen violent and bloody turbulence between human groupings. The massacre of Tutsis by Hutus resulted in some 800,000 being

Open tombs of some of the 500,000 murdered Tutsi people at the Rwandan Genocide Memorial in Kigali.

killed over a three month period after these ethnic groups in Rwanda had peacefully shared their homeland for centuries. German and later Belgian colonial administrators stoked ethnic resentment by favouring one group over the other in pursuit of their own colonial interests. So some strife between Tutsi and Hutu communities had begun to bedevil Rwanda even before its independence from Belgium in 1961.

The early 1990s saw Rwanda being ruled by a Hutu-dominated government that was fighting a civil war against Tutsi insurgents. The conflict heightened and the assassination of the Hutu president of Rwanda in April 1994 became the pretext for extremist Hutus to call for the slaughter of Tutsis and moderate Hutus. Mass slaughter was the outcome.[2]

From the 16th to the 19th century inhospitable, brutalising, warlike behaviour was an essential element in the colonisation of other lands by the peoples of Europe. In some places dispossession of previous inhabitants was achieved by genocidal means. In Australia, it was bad news for the long enduring ancient Aboriginal population when it came into contact with European colonisers. The violence and slaughter that ensued were driven by a perception of race superiority, Anglo-Saxon privilege, and a Christian religious heritage or licence that validated the domination

Mounted police kill Aborigines during the colonial period.

and civilisation of often ancient, indigenous cultures that were deemed inferior.

This historic brutality is I think a pointer to the earlier treatment meted out to Neanderthals as "modern" humans encroached into the Neanderthal European homeland.

Turning back to Australia for a moment, puzzles locked in fossilised remains of the time when the first human footprints were made on the northern coastline have been interpreted to suggest that forebearers of the original Australians arrived from the north across land and water somewhere

Gulkula, East Arnhem, Northern Territory. Aborigines' profound knowledge of animals and plants was a hallmark of an unchallenged occupancy of the continent for tens of thousands of years.

between 50,000 and 100,000 years ago. At the time, a sea journey away from Asia towards Australia, while a real adventure, may not have been so daunting to migrating people faced with rising sea levels, volcanic activity, and civil pressures. With monsoon winds blowing from the north-west, there was a fair chance that even a simple raft launched somewhere in the Indonesian island chain could have bobbed across to the coast of Australia, at times of lower sea levels when Australia was larger than it is today, and the stretches of ocean that separate it from its northern neighbours would not have been so wide or in some places may not have existed at all.

The supremacy of the Aborigines' land management skill and their sovereign reign over Australia is a remarkable chapter in the global history of indigenous peoples. It is also clear that while the changes to the biology of Australia were immense, Aborigines settled into a comfortable, harmonious existence. The development of a profound knowledge of the animals and plants of their homeland was the hallmark of an unchallenged occupancy for tens of thousands of years by these remarkable first Australians.

A major tragedy of the 19th and 20th centuries has been the desecration of Aboriginal cultures. This can be directly linked to the escalating settlement of tribal homelands and the attendant brutalisation and displacement of the original inhabitants by European colonisers. Aborigines were frequently forcibly evicted from their ancestral territory or slaughtered if they stayed.

We humans have been present on planet Earth for less than the blink of an eye, and like other species before us, we too are destined to disappear. Opinions vary about humanity precise origins. In 2019 a group of scientists claimed to have pinpointed the ancestral homeland of all humans, in Botswana around the Makgadikgadi saltpans, once a large lush lake area.[3] From there, after 70,000 years, changing rainfall patterns forced our ancestors to move up through Zambia into the Rift Valley grasslands. Eventually and from there the early Neanderthals migrated across to Europe.

Unlike dinosaurs and Neanderthals, modern humanity will become extinct by our own hand. No catastrophic event—no large asteroid, massive volcanic eruption, nor ice creeping out from the poles. With a combination of runaway population growth, resource depletion and rapid, self-induced global temperature warming, modern humanity is managing to sow the seeds of our own extinction—in perhaps just a few hundred years. We now know this and yet seem unwilling and incapable of doing anything about it. Tragically the differences from the inevitable human extinction and those that preceded it are stark. Humans will be the first species on the planet to have committed calculated suicide.

But let us back up first and consider our closest relatives in evolutionary terms—our early Neanderthal cousins. How did they emerge and set in process the march towards modern humanity?

LEFT: Neanderthal man, *Homo sapiens'* closest relative in evolutionary terms. RIGHT: Early human ancestors walking erect and human-like.

Archaeological findings show that over relatively short periods of time in a geological sense, climate change caused glaciers to advance and retreat. About three million years ago as the climate cooled, advancing glaciers and icy conditions severely impaired the habitats of species that were to be ancestors of human beings. Probably this changing climate contributed to their descent from the trees and out of forests. Many perished, but some were able to survive. When the world warmed again and the ice retreated, forests regained their former ground, even covering Iceland. At the other end of the planet, where ice and penguins are now, traces of leaf waxes in sediment cores extracted from the Ross Ice Shelf show that Antarctica once also hosted forests.[4] The margins of the Antarctic Peninsula resembled the kind of podocarp conifer and southern beech forest seen today in New Zealand's South Island and parts of southern Chile.

However, when forests advanced again, some descendants of forest-dwelling primates had adapted and now preferred grassy woodland habitats. After more than a million years of walking on two feet, legs had lengthened and opposable big toes had shortened, losing their usefulness for swinging amongst the trees, while other adaptations sharpened for survival on the ground.

Our ancestors who walked erect and human-like developed new skills and learnt to follow the fires that opened up extensive grasslands. For perhaps three million years, they were too few to occupy more than local patchworks of grassland and forests, but slowly they expanded their homelands and population.

Rock paintings attributed by some to Neanderthal "cave men". During the last Ice Age, Neanderthals sheltered in Europe's plentiful limestone caves.

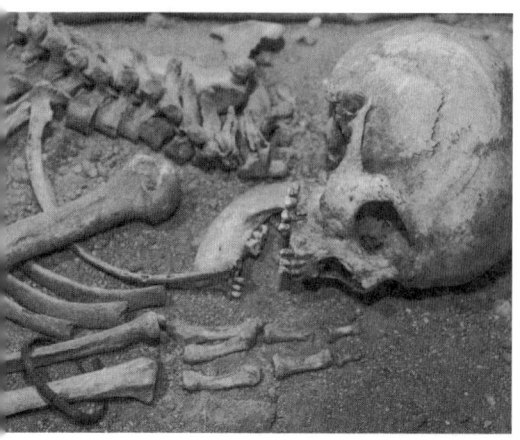

Archaeological excavations in the Neander Valley eventually determined that these skeletal remains were ancient human relatives. Many Neanderthal remains have now been unearthed.

It is thought that between 50,000 and 100,000 years ago, *Homo sapiens* possibly numbered as few as ten thousand.[5] They began to venture out of Africa, following a migratory corridor north where Israel and Palestine are today, and branched into Europe, Asia and the lands beyond.

Homo sapiens' closest relatives in an evolutionary sense, the Neanderthals, lived during the last Ice Age, taking shelter from the ice, snow, and otherwise unpleasant weather in Europe's plentiful limestone caves. Many of their fossils have been found in such caves, leading to the popular idea of them as "cavemen".

Scientific evidence suggests that modern humans venturing into Europe descended from a population of *Homo sapiens* who migrated from East Africa roughly 80,000 to 50,000 years ago and spread along the southern coast of Asia and to Oceania by about 50,000 years ago and across Europe about 40,000 years ago.[6] Neanderthals, it is thought, were thriving there when our *Homo sapiens* ancestors turned up. Some Neanderthals are thought to have ventured into Asia, and travelled as far east as Siberia.

It is thought that Neanderthals, modern humans and probably other Homo species lived alongside each other in Europe for at least thousands of years before Neanderthals died out some 30,000 years ago.

In appearance Neanderthals were similar to us, although they were shorter and had barrel chests, stocky limbs, and large noses. Their short, stocky stature was an evolutionary adaptation to conserve heat in the cold weather. Their wide noses are thought to have helped humidify and warm cold air in the frigid European weather during the ice ages. Neanderthals typically lived to be about 30 years old, though some lived longer. Estimates put the peak Neanderthal population at around 70,000.

It was the Neander Valley near Dusseldorf in Germany—where

remnants of skeletons were discovered in a limestone cave in 1856—that gave its name to Neanderthals. Since then, Neanderthal remains have been unearthed in many other places. The publication in 1859, and subsequent popularisation of Charles Darwin's *Origin of Species* encouraged palaeontologists to seek evolutionary explanations for their discoveries.

Archaeologists have discovered that Neanderthals lived in nuclear families, and believe that they cared for their sick and elderly. They also buried their dead. It is not known if they had language, though the large size of their brains makes that a possibility.[7]

Neanderthals used stone tools similar to tools used by early *Homo sapiens*, including blades and scrapers made from stone flakes. They are thought to have created tools of greater complexity as well, utilising bones and antlers.

Neanderthals may even have built boats and sailed on the Mediterranean.[8] Scientists also suggest they knew about the medicinal qualities of certain plants. Prehistoric art on Spanish cave walls—consisting of dots and crimson hand stencils—dating back to 41,000 years ago were probably the work of Neanderthal artists.

Many scientists suspect that the coming of modern humans to Europe is causally linked to the disappearance of Neanderthals. Possibly modern humans contributed to their extinction by outcompeting them for resources and through open conflict. More recent human behaviour would suggest that violent confrontation was very probable.

Another possibility is that Neanderthals didn't die out, but simply interbred with *Homo sapiens* until they were absorbed into the modern human species. Competing opinions range from a belief that the two human species definitely interbred to a denial that the two groups ever existed in Europe at the same time.

Ed Young, writing in the September 2019 edition of *National Geographic* magazine, is emphatic, saying that some of the first *Homo*

Bushmen in the Kalahari Desert, Namibia. Most humans, with the possible exception of Africans whose ancestors never left Africa, are part Neanderthal.

sapiens to leave Africa mated with Neanderthals and produced fertile offspring.[9] He claims that as a result, all non-Africans today carry a small account of Neanderthal DNA. Those genes may boost their immune systems and vitamin D levels, but also their risk of schizophrenia—and excessive belly fat! Not your problem?

There is genetic evidence to support the interbreeding theory. In 2010, a team of scientists comparing a rough draft of the Neanderthal genome with that of modern humans concluded that most humans have one to three per cent Neanderthal DNA. The team suggested that the first opportunity for interbreeding of the two human species probably occurred about 60,000 years ago after *Homo sapiens* had left Africa, but before they had made significant inroads into Europe.[10] However, the scientists acknowledge a possibility that interbreeding may have begun as recently as 37,000 years ago. According to this idea, humans today— with the possible exception of some Africans, descended entirely from ancestors who never left Africa—are part Neanderthal.

Some researchers reckon it was climate change that led to the demise of Neanderthals. We do know that in prehistoric times the climate repeatedly changed across Europe. Sometimes these changes were relatively rapid, but from the perspective of a human lifetime, nature's changes are always imperceptibly slow. Changes in global weather patterns were associated with fluctuating sea levels and these oscillations, as ice ages came and went, had a profound impact on the land. No doubt Neanderthals would have been caught up and adversely impacted by such changes.

At the end of each ice age, when sea levels rose and polar ice caps retreated, low-lying coastal areas would have been submerged under the encroaching sea and inhabitants occupying this territory would have been displaced. We are now confronting a similar reality, likely to occur over just a few hundred, rather than thousands of years.

So down from the trees and out of the forests of Africa, then wandering across Europe to Asia, Australia and beyond, an insignificant biped once numbering just a few thousand has exploded to today's billions, dominating the global environment and fellow travellers on spaceship Earth, and possibly pushing itself to the very edge of extinction.

PLANET EARTH—
WE HAVE A
PROBLEM

Too many people, but what to do about it?

It took hundreds of thousands of years for humanity to reach a population of ten million, and that was about 10,000 years ago. The reason the population remained low and steady until relatively recent human history was that people died about as fast as others were born. For tens of thousands of years, birth rates might have been high, but so were infant mortality rates—many didn't live to see their first birthday.

The fact that population grew slowly meant that the average number of children in a family who lived long enough to have children of their own was barely more than two. For every family with more than two who survived to adulthood the arithmetic suggests others had just one, or none who survived. At any number below two, the population contracted.

Human numbers had grown to only a few hundred million about 2000 years ago, and to 2.5 billion by 1950. Within less than the span of a single lifetime, numbers more than doubled to 5.5 billion by 1993.[12] We now face the prospect of a further doubling of the population within the next half century.[3] How to address this population explosion will be a critical issue if humanity is going to survive. So, population trends can fairly be described as the elephant in the room when it comes to reasons behind escalating resource consumption and environmental deterioration.

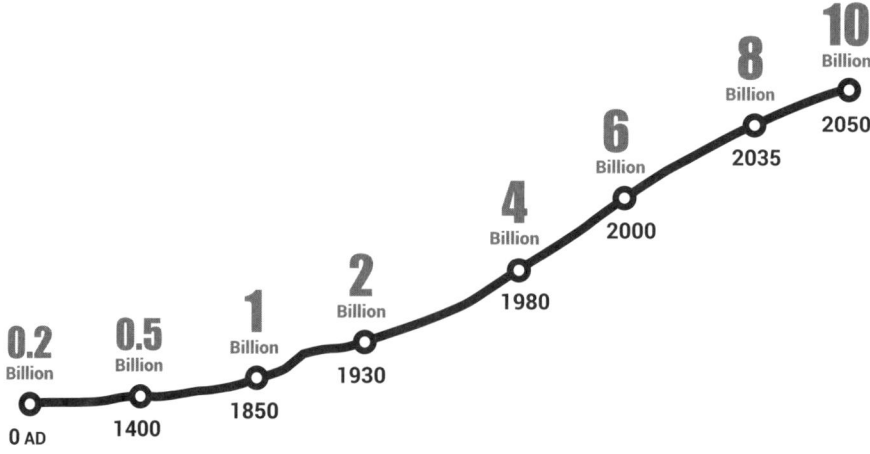

It took hundreds of thousands of years to reach a human population of ten million. Population grew to over seven billion in mid-2012 and is headed to ten billion by 2050.

Since civilisation began, decisions based on politics and military manoeuvring, along with business and cultural pressures, have determined population trends. Religions have played and still play an unfortunate role in perpetuating climbing populations and impeding rational measures to rein in runaway population growth.

Women devotees participate in a festival through the streets of Attukai, India. Population trends can be labelled as the elephant in the room when it comes to reasons behind escalating resource consumption and environmental deterioration.

Throughout history Christian churches have attempted to obstruct progressive moves by society. They have steadfastly opposed women's rights, contraception, abortion and marriage equality—the list goes on.

In the Philippines, one of the poorest countries in Asia, where the population grows by two million people each year, the Catholic Church still wields significant power and influence. The Church strenuously opposed the Reproductive Health Bill, with Philippine bishops vowing to defeat every politician who supports the legislation, to fire those at Catholic universities believed to favour the bill, and to appeal to the Supreme Court to declare it unconstitutional.

In mid-2012 the world's population had grown to over seven billion. Developing countries accounted for 97 per cent of this population growth due to their high birth rates and young populations. The radically different demographic situation between developed and developing countries illustrates the substantial gulf in birth and death rates. On one side are mostly poor

Catholic cardinals from all over the world at the funeral of Pope John Paul II at the Vatican. Through most of its history the church has obstructed progressive moves by society.

countries with relatively high birth rates and low life expectancies. On the other side are mostly wealthy countries where birth rates are now so low that population decline is all but guaranteed, and where life expectancy extends past age 75, creating rapidly ageing populations.

Conversely, in the developed countries the annual number of births barely exceeds deaths, and average population ages have climbed steeply.

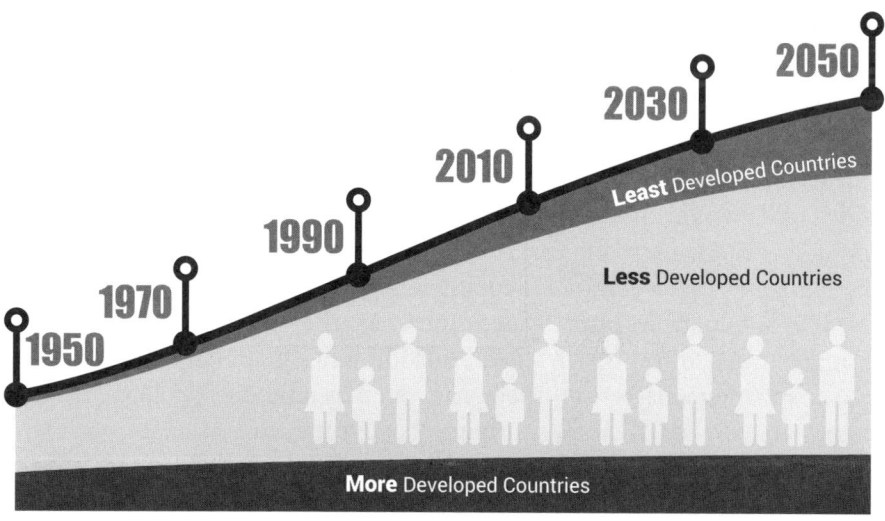

Data confirms a continuation of long term global demographic trends and a larger global population. Developing countries accounted for 97% of this growth.

By 2025, it is likely that deaths will exceed births in developed countries.[4] This decline is in large measure the consequence of near universal reduction of fertility—women are marrying later or not at all, postponing childbearing and having fewer or no children.

While virtually all future population growth will be in developing countries, the poorest of these countries will see the greatest percentage increase. Such countries have low incomes, high economic vulnerability, and poor human development indicators, such as low life expectancy, very low per capita income, and low levels of education. Thirty three of these countries are in sub-Saharan Africa, including Burundi, Ethiopia, Mozambique, and Zambia; 14 are in Asia, and one in the Caribbean. Their population is growing at 2.4 per cent per year and projected at least to double to 2.3 billion by 2050.[5]

With a current population of about 5 billion, Asia will likely experience a much smaller proportional increase than Africa, but will still add about a further billion people by 2050. Asia's population growth will largely be determined by what happens in China and India, which together account for about 60 per cent of the Asian population. Several of the more economically advanced Asian countries, including Japan, Singapore, South Korea, and Taiwan have low birth rates. In Japan, 24 per cent of the population is already aged 65 and older, a proportion certain to continue growing.

Pakistan is a scary illustration of the likely consequences of out-of-control population growth. Within the next two decades Pakistan will pass Indonesia as the most populous Muslim nation.[6] Indonesia, with 250 million people, has one of the developing world's better family-planning programs; still it will add 40 million by 2030. Pakistan, currently with three-fourths of Indonesia's population, is likely to double in numbers by mid-century with a projected 395 million people—more than today's United States of America—living in a land the size of France or Texas.[7]

Today, 60 per cent of those Pakistani millions are under thirty with one-third of Pakistani children chronically malnourished. Hunger and double-digit unemployment increase along with the population. Unemployed young men grow frustrated, and angry. A nation filled with angry young men is not a stable place, especially when they are tempted with paid opportunities to commit mayhem, including international terrorism.

Young men with few prospects, powered by religious ideology—gun in hand—have, and will likely continue to focus their frustrations on perceived wealthy oppressors. Despite the military superiority of the United States of America and its allies, individual deaths seem inconsequential and battalions of angry young men will keep on coming—they have nothing better to do and nothing to lose.

Although some growth rates will fall, the annual increase in world population will remain high; 57 million a year on average between 2000 and 2050. This is fewer than the 71 million people added annually between 1950 and 2000, but over 50 years the increase will still be more than twice the current population of China— and more than twice the current combined population of all the regions traditionally classified as developed.[8]

Powered by religious ideology young men with few prospects and gun in hand focus their frustrations on perceived wealthy oppressors. They have nothing better to do and nothing to lose.

Population growth is adversely impacting many aspects of human welfare. Nearly all our economic, social and political problems are much more challenging to solve in the face of runaway population growth.

While human population quadrupled over the past hundred years, consumption of resources, as measured by combined gross domestic products worldwide has increased by a factor of seventeen.[9] This feasting at the planetary buffet table has been enjoyed by a comparative few, at the expense of many. An unequal distribution of goods, which caused woes and wars even in biblical times, has perhaps never been so skewed as it is today. Presently in terms of resource consumption the richest fifth of the world's people take 66 times as much as the poorest fifth.[10]

Can we, as a human society, make it widely understood that failure to reduce population is likely to lead to a population crash when fossil fuels, fresh water and other resources become scarce and civil strife erupts? This, the 21st century, will likely be the century that determines what the human carrying capacity is for our planet. It will come about in one of two ways. Either we get on the front foot and decide to manage our own numbers, or nature will do it for us, in the form of climate chaos, famines, thirst, crashing ecosystems, opportunistic disease and conflict over dwindling resources.

Is there an acceptable, non-violent way to convince people of all cultures, religions, nationalities, tribes, and political systems that it's

Volunteers hand out food to pilgrims in Lalibela, Ethiopia. The twenty first century will likely be the century that determines the human carrying capacity of the planet.

in their best interest to manage population levels? China has attempted it, but the idea conjures frightening images of coercive governments invading our homes, and perhaps legitimising state-sanctioned genocide. Yet some cultures have found non-intrusive ways to encourage people to have smaller families, manage population levels and related demographic challenges.

With food production currently occupying about 40 per cent of the Earth's non-frozen terrestrial surface, plus infrastructure, cities, and towns, we've claimed nearly half the planet for just one species—us. What about all the other species? Not surprisingly many are on the brink of extinction.

Tiger … survival threatened by poaching and habitat loss. With less than 4,000 now thought to exist in the wild, the tiger joins many animals on the brink of extinction.

I reckon the medical profession has a lot to answer for in terms of population expansion. Until relatively recently, as we have noted, population increase was pedestrian. Occasionally populations contracted dramatically, such as during the Black Plague, that killed off an estimated one fourth of humanity across Europe in the mid-14th century. However, things started to change with the introduction of vaccines. In 1796 British Surgeon Edward Jenner discovered a vaccine for smallpox, a disease that used to knock back populations by the millions each year. Jenner's cure inspired 19th century French chemist Louis Pasteur to develop others, against rabies and anthrax.

Over the past two centuries, we have become brilliant at beating back diseases or pre-emptively protecting ourselves from them. Through much of the world, we've doubled average human lifespans from under forty years to nearly eighty.

Pasteur made two other key contributions to human survival and population expansion; the first was the now familiar dairying process of pasteurisation. Pasteurisation extended the shelf life of milk, which improved nutrition and reduced infections from pathogens such as salmonella and those which cause scarlet fever, diphtheria, and tuberculosis.

Pasteur was one of those who was instrumental in convincing humanity that infection does not spread by some spontaneous magic, but is caused by germs. In the 19th century, for the first time, hand soap became common in homes and hospitals. Before then patients died as

LEFT: Edward Jenner (1749–1823) discovered a vaccine for smallpox, a disease that used to knock back populations by the millions each year. RIGHT: French chemist and microbiologist Louis Pasteur (1822–1895) developed vaccines for rabies and anthrax.

Early European settlers found the bark of the cinchona tree could both cure and prevent malaria or ague, endemic in Europe at the time.

often from infections picked up from a surgeon's unsterilised hands and scalpels as from the ailment he was trying to rectify.

One of the first uses of surgical disinfectant was in a maternity ward in Vienna,[11] where doctors, by washing their hands in a chlorine solution, lowered both baby and mother mortality by a factor of ten—an innovation with a very direct impact on the number of living humans.

Advances in medical science are heralded as great progress—triumphs of human intelligence that are extending life spans and perpetuating population growth. Had we let nature take its usual course most of the 2.3 billion of us over forty would not be around—and that includes me! Almost half of all children would have died before age five and at least a fifth of all women would have died during pregnancy or from childbirth complications.

What about something of the contribution trees have made to the advancement of medicine? Cinchona, now the national tree of Peru and Ecuador, changed the course of history.[12] There are more than 20 species of this impressive 25 metre tree with large, shiny, conspicuously veined leaves and deliciously fragrant flowers, but its real claim to fame is the effectiveness of its bark for treating malaria.

In the early 17th century, when Spanish colonists and Jesuit missionaries in Peru were introduced to cinchona bark, there was no malaria in South America. Cinchona was likely an indigenous Quechua medicine used to treat an unrelated fever, and it was this that inspired Europeans to make an incredibly lucky guess. In Europe where malaria or "ague" was endemic, they found that cinchona bark could both cure and prevent the disease.

The baptism of Louis de France, son of Louis XIV. Cinchona cured him, and was soon widely accepted as the only preventative and cure for malaria.

Its reputation and use spread quickly through Spain. Ironically, it was probably the Spanish, via their African slave trade, who brought malaria to the one continent that didn't have the disease, but had the cure for it. In a controlling partnership with the Quechua, massive logging of cinchona trees began, with fleets of ships carrying the bark back to Europe.

"Jesuit bark" was regarded with suspicion by Protestants because of the connection to Spain and Catholicism. In England Oliver Cromwell died of malarial complications rather than take the "powder of the Devil". But in 1679 cinchona cured the son of Louis XIV of France, and for more than 250 years it was widely accepted as the only preventative and cure for malaria, until alternatives were synthesised.

Cinchona bark contains a cocktail of alkaloid drugs—probably evolved by the tree as a defence against insects—that would indeed have been medically valuable to the Quechua. The name *quinine* came from the Quechua language, *quina quina*, "the bark of all bark". Quinine alkaloids have the rare capability of making certain components of our blood poisonous to the malaria parasite.

Malaria was a problem in Europe until the 20th century, but in the tropics the disease was the limiting factor to European colonial ambitions. Virulent strains killed more than half of all Europeans who ventured to some parts of Africa and Asia.

In the British settlements of Virginia, more people died from "swamp fever", as it was known there, than at the hands of the native Americans.

Anything that might check the disease was of the upmost strategic significance and commanded a high price. To protect their lucrative monopoly, the countries of South America imposed a death sentence on anyone exporting cuttings or seeds. However, their forests could not meet the voracious demand for quinine, and in the 19th century the Dutch and British managed to smuggle out cinchona seeds to start their own plantations.

Without cinchona, European colonial empires would not have been expanded so much in tropical parts of the world. In India, the British Raj depended on quinine, the white powder extracted from the bark was taken daily in "tonic water". Gin, lemon and sugar were added to make it more palatable, resulting in today's Gin and Tonic. Modern tonic water has much less quinine—enough, though, to make it fluoresce pale blue under the glow of ultraviolet nightclub lights.

In the 20th century, medical advances kept coming, each saving—and extending—more and more human lives. After Cuban microbiologist Carlos Finlay pinpointed the carrier of the yellow fever virus, American doctors William Gorgas and Walter Reed implemented the world's first massive mosquito control programme,[13] without which the Panama Canal would never have been completed. Vaccines for diphtheria, tetanus and polio, and the crucial invention of antibiotics, all lowered mortality and increased longevity—which meant that more people, young and old, lived.

Australian pharmacologist and pathologist Howard Florey made a substantial contribution to this transformation. Born in 1898 Florey

Cuban microbiologist Carlos Finlay (on a 1993 Cuban stamp), who pinpointed the carrier of the yellow fever virus.

shared the Nobel Prize in Medicine in 1945 with Ernst Chain and Alexander Fleming for the development of penicillin. Florey carried out the first ever clinical trials of penicillin in 1941. Subsequently he and Chain made a useful and effective drug out of penicillin, after others had abandoned the task as too difficult. Working out how to turn penicillin into a medicine saved hundreds of millions of lives. Sir Robert Menzies, Australia's longest-serving Prime Minister said, "In terms of world well-being, Florey was the most important man ever born in Australia."[14]

With explosive speed, humanity has become a force like no other species in the 3.8 billion years of life on Earth. Let's be really honest here—it's people, or more the excessive number of people, that are at the centre of the world's pressing challenges. A finite planet and too many people—and the numbers are growing.

"We are the most numerous mammal on the planet," says David Suzuki, "and our numbers and longevity alone mean that our ecological footprint is huge: it takes a lot of air, land and water to meet our basic needs."[15]

Of course, the multiplication of humanity is bad news for the climate. Expanding populations need controlling and the lack of government commitment and political courage to do so is likely to continue. This raises some really challenging questions like how much ecosystem is required to maintain human life? At what point does our overwhelming presence displace so many other species that eventually we push something off the planet that we didn't realise our own existence depended on?

Human skulls from the Killing Fields at Choeung Ek Genocidal Centre, the site of a former orchard and mass grave of victims of the Khmer Rouge. The killings in Cambodia between 1975 and 1979 parallel other genocidal dark chapters in history.

The notion of husbanding humans as though we were chickens or livestock horrifies on moral, religious, and philosophical, not to mention legal grounds. To suggest applying principles of wildlife management to our own species conjures abominations such as humans being culled like so many pest species. Such suggestions quickly draw parallels with dark chapters in history of attempts to thin human ranks—otherwise known as genocide—that remain among our most indelible memories.

Except for volcanic eruptions and earthquakes, every emergency on Earth is now either related to, or aggravated by the presence of more people than conditions can reasonably tolerate. We must hope that humane and non-violent means of gradually bringing our numbers down emerge. The alternative is letting nature do it for us—and that is unlikely to be gentle!

If we stop reproducing completely, in little more than a hundred years our population would be zero. Let's suppose, putting the myriad objections aside, that the entire world adopts a one-child policy immediately. By the end of this century, we would be back to 1.6 billion, the same population as 1900. Holding to just one offspring per family for a few generations would exponentially bring us back to an ecological balance with the Earth's carrying capacity. Worth thinking about.

NATURE'S LIGHT SHOW

From green bacteria big things grow

Hundreds of millions of years before our earliest primitive ancestors came down out of the trees, life evolved within a thin, fragile envelope of soil, water and air we call the biosphere. Today the survival of all the planet's plant and animal life remains absolutely dependent on the continuing efficient functioning of this biosphere.

From other planets in the Solar System we can tell almost certainly that Earth's original atmosphere was a reducing one. What does that mean? The atmosphere was in a condition in which the process of oxidation was prevented by the absence of free oxygen. Nowadays iron can react with oxygen to produce rust, iron that has been oxidised. The iron has lost some electrons and the oxygen has been "reduced" by gaining some electrons and is no longer available in the free form for further oxidation.

When oxygen first appeared in the atmosphere it was a poison.[1] Paradoxical though this reality sounds, some suggest oxygen is still a poison: you can go to your local chemist and buy "anti-oxidant" dietary supplements, said to improve health and limit diseases. Antioxidants inhibit oxidation that may damage cells in the body.

When oxygen atoms first became present the atmosphere, they were not free as oxygen gas, but tied up with other elements in compounds such as carbon dioxide and water. Free oxygen was a later arrival. Oxygen was then a polluting waste product of green bacteria. Primitive organisms

Continuing survival of the Earth's plant and animal life remains absolutely dependent on the efficient functioning of the biosphere.

evolved the ability to cope with this oxygen. Over countless generations, living things then came to depend on it.

All the free oxygen in the atmosphere comes from green bacteria—some of them free living, some in the form of chloroplasts, which we will get to over the page. This triumph of evolution is a neat and self-sustaining cycle that forms the basis of life on Earth.

Except for a few organisms able to draw energy from hot deep ocean vents, all life on Earth depends either directly or indirectly on energy generated by the Sun. So the foundation of evolving life is plants, in particular, green tissue that converts carbon dioxide and water into carbon-based sugars which store the Sun's energy.

Ancient cycads belonging to an early plant group now found in the tropics. Before animals evolved primitive plants could already convert carbon dioxide and water into carbon-based sugars.

At present, significant quantities of carbon are locked up in living bodies, both plant and animal. A far greater volume of carbon is stored away in rocks, such as chalk, limestone and coal, which come from the remains of living bodies. In the early evolution of the planet those same carbon atoms would mostly have been in the atmosphere as compound gases, such as carbon dioxide and methane.

Today carbon dioxide forms just three to four parts per ten thousand of the gaseous composition of the Earth's atmosphere, but it plays a critical role in the balance necessary to all life. Mars and Venus are considered to

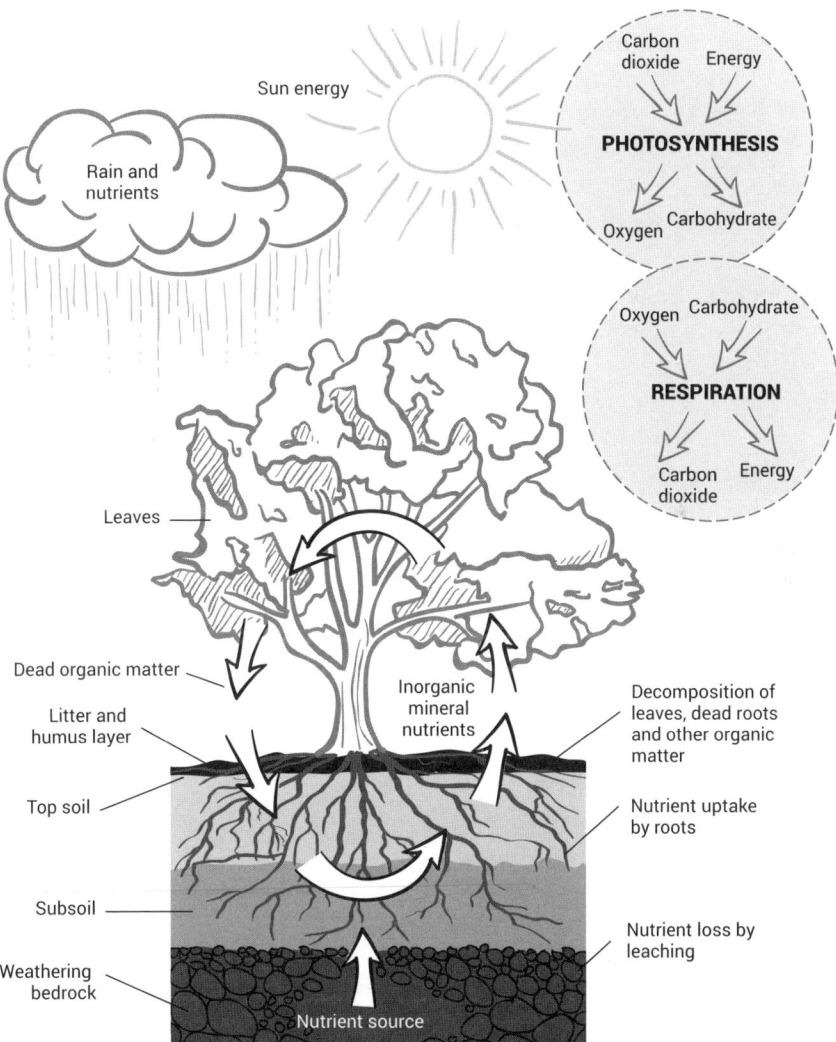

Photosynthesis and mineral cycle. The process of photosynthesis results in the production of oxygen and the absorption of carbon dioxide.

be "dead planets" where carbon dioxide makes up most of the atmosphere, as it would do on Earth if plants and the Earth's processes did not keep it within bounds.[2]

Humans and other animals have become dependent on oxygen, via the biochemical wizardry of oxygen-loving bacteria called mitochondria. Mitochondria within a cell have the function of generating most of the cell's supply of adenosine triphosphate, used as a source of chemical energy. They are referred to as the powerhouse of the cell.

Within the green, thin and vein-covered fabric of leaves the miraculous process of photosynthesis provides the spark of life. Leaves turn carbon dioxide from the air into giant trees and other plants. Within these green leaf factories, carbon dioxide we and other animals breathe out is combined with water drawn up from the ground into carbon-based sugars and cellulose to make wood.

The green colour of trees, shrubs, grasses and algae comes from the small green bodies within their cells called chloroplasts. Chloroplasts are where the work of capturing the Sun's energy and converting it to sugars takes place. Because chloroplasts are distant descendants of those free-living green bacteria we have discussed, they still have their own DNA. At a very early stage in plant evolution, green bacteria infected plant cells, and ever since, their descendants have multiplied within cells by asexual division, building up to a substantial population inside the host cells.

Trees grow from the air by photosynthesis—drawing in carbon dioxide through tiny holes in their leaves call stomata. A forest is really nature's gigantic carbon dioxide vacuum cleaner that constantly filters out and stores this component of the air. The biological alchemy is powered by energy from sunlight. Combining carbon dioxide from the atmosphere, and water from the soil, plants create chains and rings of carbon molecules that are the backbone of the web of all living things.

Plant chloroplast cells under the microscope.

Photosynthesis is Greek for light synthesis. Chloroplasts in the

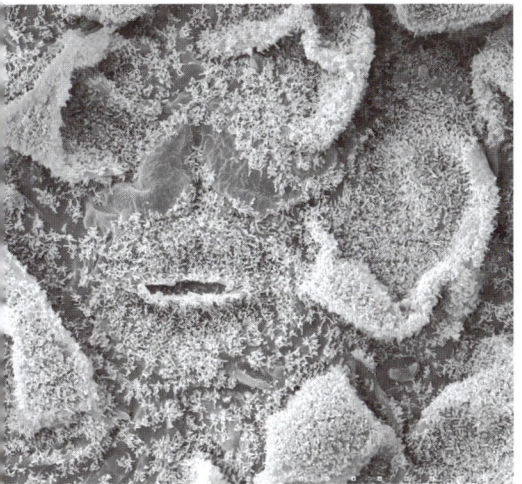

LEFT: Microscopic view of leaf underside stomata. The apertures open and close, absorbing carbon dioxide and releasing oxygen and water vapour. RIGHT: Chlorophyll has a shortcoming, a so-called green gap. Because it cannot use this part of the colour spectrum, it is reflected back unused, so leaves look green.

leaves, and other green parts of the plant, use their green pigment to trap photons from the Sun and channel the energy in the useful direction of synthesising organic compounds. The oxygen "waste" is partly used by the plant and partly exhaled into the atmosphere through the stomata. The organic compounds synthesised by the chloroplasts are ultimately made available to host plant cells.

But why are plants green and the world full of colour? Sunlight is white, and when it is reflected, it is still white. We see all sorts of colours because every object absorbs light differently or converts it into other kinds of radiation. Only the wavelengths that remain are refracted and reach our eyes. Therefore, the colour of organisms and objects is dictated by the colour of the reflected light. When rays of sunlight reach leaves, light rays with wavelengths in the green spectrum bounce back—so we see green. The other light wave lengths—reds, blues, indigoes and violets—are trapped in the chloroplasts where their energy is captured to break apart molecules of carbon dioxide in the production of sugars.

If trees, or more precisely their chlorophyll, processed light from the whole spectrum, there would be hardly any left over—and the forest would look as dark during the day as it does at night. However, chlorophyll has one shortcoming, a so-called green gap, and because it cannot use this part of the colour spectrum, it has to reflect it back unused. What we are really seeing is waste light, the rejected part that trees cannot use, and that's why leaves look green to us.

Simple sugars produced by photosynthesis are combined into larger, more complex cellulose or lignin sugars, or stored as energy reserves in starches and tubers. It is from these complex sugars that the plants themselves are built, and animals and indeed civilisation.

🌳 🌳 🌳

Two chemical processes operate simultaneously in plants. Photosynthesis converts carbon dioxide and water into a carbon-based sugars, and oxygen is released as a by-product. Respiration, or oxidation, the other chemical process, reverses photosynthesis and converts sugar and oxygen back to carbon dioxide and water. The oxidation reaction mechanism is used by both plants and animals to release the energy they need for growth, activity and reproduction.

So long as the carbon-based molecules created by photosynthesis are stored in the structures of trees and other plants, then that carbon remains bound up, or sequestered, and out of the atmosphere. Clearly, the protection and preservation of photosynthetic activity must be an important consideration in offsetting the impact of climate change.

Yes, plants give off oxygen, but after a plant dies, the chemical reactions of decay are equivalent to burning all the carbon-based materials that it has manufactured and stored. Those reactions would use up an amount of oxygen equal to all the oxygen released by that plant during its lifetime. When fungi and bacteria break down the wood, process the carbon dioxide, and breathe it out again, the same quantity of greenhouse gases is released as the plant used up to make the wood. That's why burning wood is climate neutral. It makes no difference to the volume of greenhouse gas released whether it's small organisms, or home or industrial fire that takes on this task.

If all the fossil fuels in the world were burned so much of the oxygen in the atmosphere would be replaced by carbon dioxide that the planet would go back to the status quo that applied before oxygen breathing animals could survive. We should not forget that the only reason we have oxygen to breathe is that most of the carbon in the world is tied up underground. We dig it up and burn it all at our peril.

GETTING TO GRIPS WITH TREES

Feats of architectural and engineering excellence

I think there is something quite sobering about standing next to a really big tree. Knowing that it is hundreds of years old somehow puts your human life into perspective.

Trees and human beings have long had a close working relationship, and you would have worked out from the title that this book is heavily into trees. So what is a tree, exactly? Probably most would agree that the basic features of a tree are that it is tall and long lived. Trees are simply plants that have learnt to grow up high and live for a long time. They grow tall to compete with other trees—racing upwards and spreading outwards for sunlight and water. Once up there in the sky, except for other trees, a tree has no competitors to worry about. Usually a tree has a rigid, woody, strong, expanding trunk, encased and protected by a layer of bark. The trunk supports a crown of branches, twigs and leaves. A complex root system anchors the tree to the ground and absorbs water and nutrients.

Trees have won the competition for sunlight by growing tall trunks. It stands to reason that a plant that grows a long sturdy trunk must live for a very long time, storing an enormous amount of energy in its wood. To grow its trunk, a mature tree needs as much sugar and cellulose as there is in a hectare of crops like wheat or barley. It takes decades to grow an impressive, tall trunk and the associated superstructure.

Less obvious perhaps is that trees grow from their top. Some other tall plants, such as palms and bananas, actually grow from their base—from the ground. They are not trees.

Trees come in an amazing variety of forms from tall and narrow—as are many conifers—to broad and spreading, like European oaks and African umbrella thorn trees. A tree's height and shape are determined by both its genetic blueprint and its environment.

Responding to the constant imperative of seeking resources from the sky and the soil, trees can become huge—they just keep on growing. Trees are the biggest organisms that have ever lived, with some giant specimens ten times heavier than a full-grown blue whale. Trees are incredibly diverse, with many thousands of species living in a wide range of habitats.

The list of impressive credentials of trees goes on and on. They are also the longest living organisms on the planet. Many live hundreds of years and some thousands of years. The bristlecone pine can live for more than 4,500 years on the cold, dry mountains of the United States of America, from the Mexican border north to Colorado. California's redwood and Australia's mountain ash, huon pine and messmate are also among the world's longest-lived trees.

In the big tree stakes the California redwood and New Zealand's kauri have to be included. Mature kauris are huge trees. Giants with tall cylindrical trunks topped by enormous spreading crowns. Some of them are two thousand years, even three thousand years old. Kauris with their long, minimally tapering trunks and exceptional height and girth yield the greatest trunk volume of any tree.[1]

Australian mountain ash forest, Victoria. Trees have won the competition for the Sun because they grow tall trunks. To grow its trunk, a mature tree needs as much sugar and cellulose as there is in a hectare of crops.

African savannah, Serengeti, Tanzania. Trees come in an amazing variety of forms from tall and narrow – like many conifers – to broad and spreading like European oaks or these African umbrella thorn trees.

So how long can a tree actually live? The oldest trees are neither the bristlecone pines nor kauris, but ancient spruce trees in Dalarna province in Sweden. Researchers have identified an old spruce in Dalarna with a carpet of flat shrubby growth around its single trunk. All this growth belongs to one tree. Its roots have been tested, using carbon 14 dating.[2] Research has revealed the Dalarna spruce to be an incredible 9,550 years old.[3] Individual shoots were younger, but these new growths from the past few centuries were not considered to be stand-alone trees, but part of

the original larger whole. Research on the Dalarna spruce has thrown a number of scientific schools of thought overboard. No one had any idea that spruce could live much more than five hundred years. The research also demonstrates that what grows underground is the most permanent part of the tree.

Tāne Mahuta, a giant kauri tree in Waipoua Forest, Northland, New Zealand is the largest kauri known. Some trees are 2,000 years, even 3,000 years old with tall cylindrical trucks topped by enormous spreading crowns.
Source: John Halkett

Clever they may be, but we know trees can't walk. They can still cover considerable distances given enough time. Without feet or wheels, their answer lies in the transition from one generation to the next. A single tree, that has to stay where it puts down roots as a seedling, can reproduce, and tree embryos packed into seeds are free to move. Move they often do! The moment seeds fall from the tree, their journeys begin—some short and some over considerable distances.

Some species can travel quickly by equipping their seed offspring with fine hairs so they can drift off on the wind. Poplars, willows and other species that rely on this strategy have light seeds that float away on the breezes on journeys that may cover many kilometres. Other species have a clever winged design for their seeds, which slows the seeds as they fall so that if the wind is blowing it can carry the seeds for serious distances.

Species that produce seeds enclosed in heavy fruit, such as oaks, chestnuts and beeches can't travel any distance at all, so they enter into an alliance with the animals that eat the fruit and accidentally distribute the seeds. Rodents and squirrels make a contribution to the next generation of trees. Because these small creatures often bury their winter stores of fruits and nuts a little distance away from the parent tree, the species creeps across the landscape.

Trees need to be mobile because the climate is always changing, very slowly over the course of many thousands and thousands of years. Whatever built-in tolerance trees have, if it becomes too warm, too cold, too dry or too wet for a particular species of tree at some place and time, the species must pack up and move to a more favourable location.

Bristlecone pines on the cold, dry mountains from the Mexican border north to Colorado in the USA, can live for more than 4,500 years.

LEFT: Maple tree seeds. Some species have a clever winged shape, which slows the seeds as they fall. If the wind is blowing, seeds travel a long way. RIGHT: A red squirrel. Rodents and squirrels contribute to the next generation of trees, often burying nuts a little distance away from the parent tree, so the species creeps across the landscape.

Ice ages have had a huge influence on the travel itineraries of trees. As the centuries get increasingly cold, trees move towards the Equator. Slowly over many generations trees in central Europe successfully relocated to the Mediterranean region, and in the Southern Hemisphere broadleaf tree species supplanted once-dominant early conifer species. Tree species that can't adapt to gradual climate change, whether it be a cooling as ice advances or warming as ice retreats, become less competitive and are crowded out by those species more able to adapt and move.

Recently I looked at evidence of this shifting tree phenomenon in Antarctica. With a surface area in excess of 14 million square kilometres Antarctica is larger than Europe, and almost twice the size of Australia. As much as 98 per cent of the land's surface is covered in thick, compacted ice, on average over two kilometres deep. Scientific discoveries suggest that this continent—sometimes nicknamed the *Great White Desert*—used to host some forest cover.

Leaf waxes found in sediment cores extracted from the Ross Ice Shelf[4] are evidence establishing that the Antarctic coast was once lined with beech and conifer forest. A period of warmer climate around 15 million years ago may have resulted in areas of the Antarctic resembling the kind of forested tundra seen today in New Zealand and parts of Chile.

Researchers found that concentrations of pollen and wax from 16.4 million and 15.7 million years ago represented two brief warming periods, each of about 30,000 years. Researchers could tell that the leaf wax and pollen did not blow in from elsewhere, but came from two species of trees—podocarp conifer and southern beech—that, during these warmer periods, grew around the margins of Antarctica. The presence of these species illustrates just how much warmer Antarctica was in the past and what could happen in the future as global warming rapidly advances.

From samples of fossilised wood found on the Antarctica Peninsula, Patricia Ryberg at the University of Kansas Biodiversity Institute concluded that much of the ring structure shared characteristics with the wood of tropical trees.[5] As tropical trees experience less of a seasonal effect, but are known to go through periods of short-term dormancy, this might well account for how the forests of Antarctica were able to survive during extended periods of darkness when light required for photosynthesis was unavailable for months at a time.

Beech forest, southern Chile. Evidence of leaf waxes found in sediment cores extracted from the Ross Ice Shelf shows that the Antarctic coast was once lined with beech and conifer forest similar to forested tundra today in New Zealand and parts of Chile. Source: John Halkett

AT THE CENTRE
OF HUMAN
CONSCIOUSNESS

Relationships between trees and humanity feature in religions

Throughout history, across the spectrum of faiths, ideas and cultures, trees have marked the boundary between human understanding and divine mystery. Trees have been viewed as places of reverence and homes of gods. Even today when trees and forests are recognised as the source of so many products and values essential to daily life, they are still revered as special places of environmental, cultural and spiritual importance—certainly to me and I hope to you.

Evidence suggests that humanity's deep, religious relationship with trees goes way back to the Stone Age, more than six thousand years ago. Humanity's oldest faiths and deepest symbols reflect an early, primitive connection to the natural world and with trees—imprinting on human consciousness a cyclic sense of death and decay, rebirth and renewal.

This relationship between trees and humanity is a dominant feature of religions. Genesis, the first book of Jewish and Christian scriptures, states: *"And out of the ground the Lord God made every tree that is pleasant to the sight and good for food. The tree of life was also in the midst of the garden and the tree of the knowledge of good and evil."*

Schwedagon Pagoda, Yangon, Myanmar. Throughout human history trees have often marked the boundary between human understanding and divine mystery. Source: John Halkett

The Gospels of the New Testament report that when Jesus was born, the Three Wise Men came to visit, probably from somewhere in what is now in either Iraq, Iran, Saudi Arabia or Yemen. The gold, frankincense and myrrh they brought seem like strange gifts for a baby, but Christians believe that gold was associated with Kings, and that Jesus was the King of kings. Frankincense was used in incense, medicine, for embalming the dead and in worship, and showed that people would worship Jesus.

Frankincense was used in biblical times in incense, medicine, for embalming the dead and in worship. When it was presented to Jesus, according to some accounts, it was worth more than gold.

Myrrh was a perfume put on dead bodies, signifying that Jesus would suffer and die.

According to some accounts frankincense was worth more than gold, and was then the most valuable substance on Earth. It is produced by boswellia trees, *Boswellia sacra* and other closely related species. Commonly known as frankincense or olibanum trees, they are native to the arid lands of Oman, Yemen and the inhospitable mountainous region of Northern Somalia.

If a boswellia tree is wounded, or deliberately sliced, tears of white or pale yellow frankincense, an antiseptic, scented mixture of resin and water-soluble gums are exuded from special ducts, to discourage termites and other insects.

Frankincense was already a valuable item for South Arabian trade by 2500BC, when the Egyptians needed a supply for embalming their dead. They considered frankincense to be 'the sweat of the gods fallen to Earth'.

Frankincense producing boswellia trees are native to the arid lands of Oman, Yemen and the inhospitable mountainous region of Northern Somalia.

About 1500BC, in what was possibly the world's first recorded plant collecting expedition, the Egyptian Queen Hatshepsut explored the possibility of growing frankincense at home in Thebes to save the expense of having to import it. According to inscriptions on temple walls, Hatshepsut sent five galleys, each powered by 30 rowers, to the Land of Punt[1]—thought to be the Horn of Africa—to bring back incense trees that were then planted at Karnak on the banks of the upper Nile. Apparently, the trees didn't survive in Egypt, so Punt and Southern Arabia continued to be the only sources of frankincense.

Starting around 1000BC an overland Incense Route became well established, running from Southern Arabia and the Horn of Africa towards the Mediterranean and Mesopotamia. It was traversed by heavily guarded camel caravans supported by strategically placed fortresses.

Over thousands of years Babylonians, Egyptians, Jews and Greeks have all required incense for their temples. Nowadays it takes a visit the Gulf States or to temples and churches to enjoy the mildly intoxicating fragrance of concentrated frankincense, a substance that has been scraped from Boswellia trees for at least five thousand years.

The foremost tree in early culture and commerce was the olive tree. From the dawn of recorded history, the olive has enjoyed a status and prestige not shared by any other plant. In the Bible, the olive tree is acknowledged as 'the first of trees'. In the book of *Judges*, chapter 9, verse 8 is written: "The trees went forth on a time to anoint a king over them; and they said unto

TOP: Ancient olive tree acknowledged as "the first of trees" in the Bible and the uncrowned king of trees. ABOVE: Jesus prays in the Garden Gethsemane, scene of the betrayal which according to the Gospels led to his crucifixion. The name Gethsemane stems from gath-semen, which means oil press – an olive oil press.

the olive tree, Reign thou over us." The uncrowned king of trees, apart from its significance to Biblical religions, has been a benchmark in economic progress.

Olive tree cultivation played an important part in early civilisations. Archaeologists have established that the wealth and power of the Minoan kings—who ruled an advanced culture in Crete from about 2700 to 1450BC—was founded on and maintained by trade in olive oil. Their palace at Knossos is considered a monument of civilisations, due to its luxury materials, impressive architectural plan, advanced building techniques and imposing size. A prominent feature is the Room of the Olive Press, where the great jars in which olive oil was stored are still in existence.

From hard necessity the ancient Jews and other Semitic peoples were nomadic. They were always at the mercy of factors beyond their control, moving on in search of those green pastures that figured in their idiom, until a time came when they were able to cease their nomadic wanderings. It seems that in Palestine the Jews and their Semite cousins learnt to cultivate the olive tree. The shade afforded by olive groves allowed better pasture to grow than could be found in arid plains. Under this shade, with walled terraces to hold the soil, they grew wheat and barley, cucumber and melon, garlic and onion, leeks, lentils and beans.

The olive trees were venerated and the Bible makes frequent references to their importance. The Jews were so impressed by this wonderful tree, and by the blessings it had showered on them that they gave to the word *anointment* a significance in their lives which has survived through the ages. Anointment with olive oil has a symbolic significance in Christianity down to the present day.

When the Jewish people built their capital Jerusalem, surrounded by the olive groves of the Promised Land, they built it at the foot of the Mount of Olives. Nearby was the Garden of Gethsemane, traditionally the scene of betrayal that lead to Jesus Christ's crucifixion. The name Gethsemane stems from *gath-semen,* meaning oil press.

The ancient people of the Mediterranean and the Levant, who built Western civilisation—Egyptians, Cretans, Phoenicians, Hellenes, Carthaginians, Arabs and Romans—had the olive tree as a common denominator in their economy and culture.

Most of the geographic spread of the olive tree took place in pre-historic times but it is believed that it was the Greeks and Phoenicians who took the olive as far as southern France, Spain and Portugal. Large-scale olive tree cultivation did not reach North America until the 1870s when agriculture expanded in California after the gold rushes. Today, the industry is prospering with some 20,000 hectares of olive groves there.[2]

The olive tree's contribution to the development of civilisation was paralleled in more recent centuries by another tree that's contributed to a quantum leap in human development.

Native to the Amazon and Orinoco river basins of Brazil, Venezuela and Colombia, the rubber tree *Hevea bransiliensis* was first known in the Old World as *caoutchouc,* from the indigenous word *cauchy,* or 'weeping wood'. When damaged, its inner bark oozes creamy latex, a suspension in water of about 50 per cent rubber, which when exuded quickly coagulates to seal the wound. When incisions are made to tap latex, anti-coagulant chemicals have to be added to keep the latex running.

In 1531, an Aztec product caused a stir at the Spanish court when it first saw bouncy rubber balls. By the 1770s the British were using coagulated latex for rubbing out pencil marks—hence the name 'rubber'. In London little cubes of 'India rubber' fetched three shillings each—a large sum at the time.

For centuries tribes in the upper Amazon basin had been using rubber to mould shoes, and for waterproofing, long before the Scotsman Charles Macintosh[3] used dissolved rubber to treat fabric for his famous rainwear in the 1820s.

Rubber trees take between seven and ten years to deliver their first

LEFT: Rubber trees take between seven and ten years to deliver their first latex harvest. Harvesters make cuts just deep enough to tap the latex vessels without harming the tree. The latex is collected in small buckets. RIGHT: Rubber hit the road in 1888 when John Boyd Dunlop patented the first successful inflatable rubber bicycle tyre. In the early twentieth century pneumatic tyres, rubber seals, gaskets, mats and hoses for vehicles gave rise to companies such as Firestone, Goodyear, Michelin and Pirelli.

latex harvest. Harvesters make incisions across the latex vessels just deep enough to tap the vessels without harming the tree. The latex is collected in small buckets. Because latex production declines with age, rubber trees are usually felled when they reach the age of 25 to 30 years. The earlier practice was to burn them, but in recent decades the wood has been used in furniture manufacturing, especially in Malaysia.

Unfortunately, straight from the tree rubber cracks in the cold and becomes gummy in the heat. In 1839 Charles Goodyear discovered that 'cooking up' raw rubber with sulphur makes it tough and resistant to extremes of temperature.[4] This 'vulcanised' rubber appeared everywhere— in pumps, steam engines, combs and corsets.

Demand for rubber soon exceeded supply. The price rocketed, leading to a disorderly Amazon 'rubber rush' with people staking claims and over-exploiting wild trees. In 1876, Sir Henry Wickham shipped a consignment of 70,000 rubber seeds from Brazil to Kew Gardens near London. From there, seedlings were distributed to British colonies in Asia, where they were propagated with great success—the ancestors of today's huge rubber plantations.

In 1888 John Boyd Dunlop patented the first successful inflatable rubber bicycle tyre. In the early 20th century pneumatic tyres, rubber seals, gaskets, mats and hoses for vehicles gave rise to companies such as Firestone, Goodyear, Michelin and Pirelli.[5]

By the late 1930s a million tons of crude rubber was being exported annually from South East Asia, and it was the single most valuable import into the United States of America. The occupation of most rubber plantations by Japanese forces in World War II prompted the urgent development of synthetic rubber from fossil fuels and their by-products. However, half of all rubber still comes from the tree, now grown most prolifically in Thailand and Indonesia.

Now time for a drink do you think? In much of the world, an essential component of that pleasure is the cork oak,[6] a medium-sized, evergreen tree that commonly lives for more than 200 years. Cork oak is native to southwest Europe and northwest Africa and is the primary source of cork for wine bottle stoppers. The European cork industry produces 340,000 tonnes of cork a year, with a value of €1.5 billion, employing 30,000 people.[7]

The bark of cork oak is adapted to defend trees from fungi and microbes and is exceptionally impermeable even to air, and almost completely inert. No other untreated, naturally occurring plant product can remain unchanged in contact with so many substances. It is resistant to water, petrol, oil, and of course alcohol. Its cells can withstand extreme

Cork oaks commonly live for more than 200 years. Harvesting is done by hand. The European cork industry produces 340,000 tonnes a year, employing 30,000 people.

compression while retaining their springiness—perfect for squeezing tightly into the neck of a wine bottle. As a bonus, when cork is cut lots of microscopic cups are formed, and those myriad tiny vacuums prevent corks from slipping from smooth glass bottle necks.

The cork oak is rare in being so quick to regenerate its bark, which can be harvested once the tree is about 20 years old and then repeatedly about every decade. Harvesting is done entirely by hand. The bark is stripped from the trunk in late spring up to a height of about 2.5 metres and from sections of larger branches.

Cork is used to make a wide range of products, including insulation panels, floor and wall tiles and sound-proofing in the car industry, as well as for handicrafts and artistic uses such as cork paper used in printing, clothing manufacture and cork maroquinerie. Cork is also used in cricket balls, badminton shuttlecocks and handles of fishing rods. It is used for fridge insulation, engine gaskets, fishing nets, shoe heels, bulletin boards, woodwind instruments and model trains.

Cork oak trees do best in a moist maritime environment with hot summers, typical of the lower slopes of hills around the western Mediterranean. From the Atlantic coast to Italy and from Algeria to Tunisia, cork forests cover about 26,000 square kilometres although more than half of the world's cork comes from Portugal, with most of the rest from Spain.

According to Pliny the Elder, Roman women of his day appreciated the insulation and lightness of cork-soled sandals as much as the extra height it gave them.[8] In fact, having evolved to protect trees from fire, cork's thermal insulation is so good that it was used to shield the fuel tanks when NASA's space shuttles travelled to space.

At the northern end of the National Arboretum in Canberra is a hundred-year-old cork oak plantation, established many years before the Arboretum was created.

At the time, city planner Walter Burley Griffin and Charles Weston, the officer in charge of forestry for Canberra, were looking to trial different tree species. Griffin saw the potential for growing cork oaks in Canberra's dry climate, and sourced the initial acorns from the Royal Botanic Gardens in Melbourne. The acorns were propagated and planted in October 1917.[9] Demand for cork was growing and a second consignment of acorns for Canberra was gathered in Spain. However, the ship they were on was torpedoed during the First World War and the 30,000 acorns were lost at sea.

ABOVE: Harvesting balsa logs, East New Britain, PNG. Balsa trees speed of growth accounts for the lightness of the wood. Balsa is widely cultivated in the tropics where trees can grow to 25 metres with a diameter of half a metre within seven years.

RIGHT: De Havilland Mosquito aircraft. With a speed of more than 650 kilometres per hour, the "Balsa Bomber" was one of the world's fastest operational aircraft.

Another interesting tree is balsa,[10] the world's lightest hardwood. Balsa is a large, fast-growing tree native from southern Mexico to southern Brazil, which can now be found in many other countries, including Papua New Guinea, Indonesia, Malaysia, Thailand and the Solomon Islands.

Balsa trees can establish themselves in forest clearings or on abandoned agricultural fields and grow extremely rapidly. Their speed of growth accounts for the lightness of the wood, which has a lower density than cork. Trees generally do not live beyond 30 to 40 years. Balsa is widely cultivated in the tropics where trees can grow to 25 metres with a diameter of half a metre within seven years, reaching maturity in 12 to 15 years.

In living balsa trees large cells filled with water give the wood a spongy texture. Hence the wood of living trees not much lighter than water and barely able to float. When balsa wood is kiln dried the cells become hollow and empty. The large volume-to-surface ratio of the resulting thin-walled, empty cells gives the dried wood a high strength-to-weight ratio. The remaining cell structure is rigid, making seasoned balsa surprisingly stiff for its absurdly light weight.

Balsa wood is very buoyant and insulates very efficiently against heat and sound. The wood can be adapted to a great number of special uses, it glues well, stains and polishes, but is very absorbent.

Balsa wood is often used as a core material in composites like the blades of wind turbines. In table tennis bats, a balsa layer is typically sandwiched between two pieces of thin plywood. Balsa is also used in laminates together with glass-reinforced plastic, or fibreglass, for making high-quality surfboards and for the decks and topsides of many types of boats.

Balsa has also been indispensable to pioneers of the sea. Being so light it has often been used by raft builders, and indeed the Spanish word for raft is *balsa*. Christopher Columbus and other early Spanish explorers to the Americas, first encountered 'Indians' off-shore on rafts made from balsa.

Britain was short of aluminium during World War II but had plenty of woodworkers, and the de Havilland aircraft company turned to timber for its nimble Mosquito warplane with a speed of more than 650 kilometres per hour. The 'Balsa Bomber' was one of the world's fastest operational aircraft. The fuselage was made from sheets of balsa glued between layers of birch.

So just a few examples of the spiritual, historic and economic value that diversity of trees has contributed to humanity. We could fill quite a few more pages with other examples, but let's get back to matters central to the theme of this book.

FORESTS—WHERE TREES LIVE

The importance of forests stretches far beyond their boundaries

One of Australia's finest poets, the late Les Murray, had a strong native forest family connection. He wrote[1] that the "newly luxuriant woodland my grandparents John and Isabella Murray encountered when they took up their land at Bunyah in 1870 … is the forest I have known all my life, the one I walked in first and the one I'm likely to walk in most often in the future." He was a keen observer and amateur forester. "I saw the very tail end of the era of bullock teams and trundling iron-shod jinkers, always called 'trucks' by the men who used them."

In an essay he describes his affinity for his beloved local native forests: "It is a quality not so much alien and indifferent, as too many literary authors by now have parroted, but rather sober, subtle and uncorrupt, with a curious remote decency about it. As you move and work there, or as you die there, you do so in an intense spare abundance that sheds its perfumes and its high riddled light on you equally … One in which you are as much at home as a hovering native bee, or the wind, or death, or shaded trickling water."[2]

Perhaps a bit deep to start a chapter with, but if you think it through you may appreciate the sentiments.

Along with other plants, plus a host of animals, trees inhabit forests. Forests are some of the planet's great ecosystems—communities of living organisms and ecological systems interacting with the environment. Like

LEFT: Les Murray … one of Australia's finest poets. A keen observer and amateur forester. RIGHT: Australian dry eucalypt forest. Forests are one of the planet's great ecosystems – communities of living organisms and ecological systems interacting with the environment.

other key global ecosystems forests deliver life's essential ingredients— clean air, pure water, habitat and resources. Plant and animal species are the critical cogs and wheels of functioning forest ecosystems. Lose too many species from a forest ecosystem and the whole system runs the risk of ceasing to operate effectively or collapsing entirely.

Remarkably nature has endowed trees with the ability to survive in extreme weather—from sub-zero temperatures to dry scorching deserts. Trees grow in polar regions in Siberia, Alaska and Canada, where for part of the year the ground can be frozen down to ten metres.

In the Mojave Desert of California, one of the hottest and driest places on Earth, the 10-metre-tall Joshua tree survives because of its spreading, deep roots reaching down to suck up any available moisture.[3] Joshua trees survive the rigours of the desert for hundreds of years; some specimens live a thousand years or more. New plants can grow from seeds, but in some populations, new stems grow from underground rhizomes that spread out around the parent tree. The Joshua Tree National Monument is a sanctuary established in the Mojave Desert in 1936, and covers 2,250 square kilometres of scenic wilderness.

The Mojave Desert was visited by Mormon pioneers who headed into the United States' west in the mid-1800s. They took the Book of Joshua seriously: "Thou shalt follow the way," reads one of its lines, "pointed for thee by the trees." When the Mormons got to south western California, they found those remarkable giant yucca trees, their branches raised

Joshua trees, Mojave Desert, California. Ten metre tall, 1,000 year old Joshua trees survive because their spreading, deep penetrating roots reach down to suck up any available moisture.

upward as if in prayer. So, the Joshua or "praying plant" got its name.

Native Americans of the south-western United States identify the Joshua tree as a valuable resource and call it *hunuvat chiy'a* or *humwichawa*. In addition to harvesting the seeds and flower buds for food they used the leaves to weave sandals and baskets.

It is thought that trees and forests cover about four billion hectares—about 30 per cent of the world's land surface. However, the importance of forests stretches far beyond their boundaries. Forests help to regulate the planet's climate. For example, they store nearly 300 billion tonnes of carbon in their living parts—roughly 40 times the annual greenhouse gas emissions from fossil fuels.

It is estimated that between 2000 and 2017 around 15 million hectares of forests a year have been converted to other uses or lost through natural causes[4]—a scary

Trees and forests cover about four billion hectares ... about 30 per cent of the Earth's land surface. However, the importance of forests stretches far beyond their boundaries.

figure. Perhaps the more encouraging news is that annual tree plantation or forest rehabilitation is estimated at 5 million hectares and increasing.

Not surprisingly, it is mainly the presence of trees that indicates which areas are forests, woodlands, shrublands, savannas and grasslands—in descending order of tree cover or tree canopy density. However, there are large regions of the world where the march of civilisation has over the past couple of thousand years cleared large tracts of forest and other woody plant communities.

Forest cover varies enormously from one part of the world to another. It is largely climate—including summer and winter average temperatures and extremes, rainfall and its distribution over the seasons, humidity and wind—that determines height, complexity, tree density and species composition. As forest or woodland types change so too does the whole assemblage of resident animals and insects. As a rule of thumb, in a forest branches and foliage of individual trees often touch or interlock, although gaps of varying size can exist. Woodlands have a more continuously open canopy, with trees spaced further apart.

Climate and soil properties greatly influence the distribution of forest and other tree-covered landscapes. Tall forest gives way to woodland where annual rainfall generally dips below about 700 to 800 millimetres, although in some places a broad overlap zone exists.

Forest vegetation can be classified in several different and sometimes

LEFT: As a rule of thumb branches and foliage of individual trees often touch or interlock, although gaps of varying size can exist. RIGHT: Boreal forest in the Yukon Territory, Canada. The austere yet spectacular boreal forests are also known by the Russian name taiga.

confusing ways. An important attribute is forest architecture or structure, which takes in things like tree density, composition and height; the presence or absence of understory trees and shrubs, and the presence of tree-related growth forms, such as lianes, epiphytes and parasitic plants. Forests can also be classified based according to the dominant tree species, resulting in numerous different forest types, such as ponderosa pine, beech forest, oak forest and so on.

While there are too many schemes for classifying forests for us to analyse here, forests can be assigned simply to three broad climatic zones, cold, warm and hot. These zones are essentially determined by temperature, closely correlating with latitude, and replicated on both sides of the Equator, exception that cold zone forests only occur in the North. Although forests were once present on the Antarctic continent as we have reported it is now the exclusive domain of ice and penguins.

In the north of the Northern Hemisphere we have the austere yet spectacular boreal forests (*boreal* means northern). Boreal forests are also known by the Russian name *taiga*.

These forests occupy the sub-Arctic zone up beyond about 50° latitude. The boreal region encircles the Arctic across Russia, Scandinavia, Alaska and Canada. The boreal forest belt represents the world's largest land-based ecosystem, and acts as part of the largest source and filter of freshwater on the planet. Beyond the northern limit of boreal forest lies bleak treeless arctic tundra and ice.

Temperatures in boreal forests are usually extremely low, with long winter seasons. The soil freezes—only thawing for a few months in the farthest northern forests. Winter air temperatures can fall to -60°C. Soils are generally thin and nutrient poor. Most of the water is delivered in the form of snow—40 to 100 centimetres a year.

The main tree species in the northern boreal forests are evergreen conifers. Conifer means "cone bearer" and refers to the way their seeds are carried on the scales of cones rather than enclosed in a fruit developed from a flower. Boreal conifers are pines, spruces, firs and larches, adapted to very cold conditions. Their needles have a small surface area and along with a high concentration of sugars and starches, plus specialised proteins in their sap, act as a natural antifreeze. Oils and resins present in leaves, bark and wood also act as protection against cold. Dormant buds at growing tips are each coated with resin to insulate them. Some deciduous trees, such as birches, alders, poplars and willows, which shed their leaves at the end of each growing season, are also present in the boreal forest belt.

Boreal forests support the world's largest caribou population; the second-largest wolf population, and polar, black and grizzly bears. They

Grizzly bear. Boreal forests support the world's largest caribou herd; the second-largest wolf population, and polar, black and grizzly bears.

share these challenging forest environments with moose, lynxes, foxes, deer, bats, woodpeckers, and other animals able to tolerate the cold, harsh conditions.

Temperate forests occupy the zone between the boreal forests and the tropical forests of the equatorial zone. The climate is neither extremely hot nor really cold. These forests have four seasons; summer, spring, winter and autumn, plus soils that are generally rich and fertile. The summers can feel hot and dry, but in these forests, climate is rarely so harsh that the soil dries up and the plants wither and die. Likewise, the winters may produce a lot of snow, but not as severe as in boreal forest regions.

Temperate forests are found in both hemispheres from latitudes 25° to 50° in regions of north eastern Asia, North America, western and central Europe, southern South America and Australasia. They can be coniferous, deciduous or mixed depending on geography and climate.

Temperate forests include some of the world's tallest trees, such as the giant redwoods, conifers growing near the northwest coast of North America. Other major areas of temperate conifer forest include the mountains of western China, northeastern China and adjacent regions of Russia. Temperate conifer forests are also present in Japan, the mountains of central Asia and in the Himalayas, in Mexico's Sierra Madre ranges, and central Europe. The Balkans and Turkey all possess conifer dominated forests and in the more mountainous regions of the Atlas ranges of northwest Africa.

In the Southern Hemisphere there are smaller regions of conifer forests, such as the monkey puzzle tree, or *Araucaria* and the tall, long lived *Fitzroya cupressoides* conifer native to the Andes mountains of southern Chile and Argentina; the kauri and podocarp forests of New Zealand and some patches of ancient conifers in Tasmania.

Vast tracts of former temperate deciduous forest have been cleared for settlement and farming. Frequently derived from glacial deposits the underlying young soils are mostly fertile and moist. In large parts of China this clearance has been extensive; less so in Europe. In the United

LEFT: New Zealand beech forest. Temperate forests occupy the zone between the chilly northern boreal forests and the tropical forests of the equatorial zone. RIGHT: Almost all eucalypts occur naturally only in Australia. This iconic plant group is comprised of more than 700 species of trees and shrubs. BELOW: Giant redwoods along the northwest coast of North America. Temperate forests include some of the world's tallest trees.

States of America settlers have cleared vast areas of deciduous forest over the past four centuries for cities and farms.

In the Southern Hemisphere, broadleaf[5] temperate forest includes the denser forests in cooler regions of Chile and Argentina, and in Australia, New Zealand and the southern tip of Africa. Eucalypts account for more than 70 per cent of trees in Australian forests and woodlands, growing

Tropical forest, Khao Yai National Park, Thailand. The trees of tropical forests are sometimes taller than those in temperate forests, but they do not reach the gigantic dimensions of North American redwoods or the large eucalypts of Australia.

in a wide range of climates from the hot tropics, to near-desert inland plains, and up on to alpine snowfields.

There are more than 700 species of trees and shrubs, almost all occurring naturally only in Australia. Evolved from primitive rainforest ancestors, eucalypts have adapted to dry environments with nutrient-poor soils.

Many eucalypts secrete a resinous gum—hence the name by which they are known around the world—*gum trees*. They have distinctive foliage, multi-stamened flowers and seeds that are contained in woody and sometimes hard capsules, protecting them from fires and insects.[6]

Tropical forest forms a discontinuous band around the Earth, about ten degrees either side of the Equator. There is rather more tropical forest in the Northern than the Southern Hemisphere. A particular feature of tropical forest is that the overwhelming majority of vegetation is trees of almost infinite shapes and sizes. Not only are trees the dominant members, but most of the climbing plants and some of the epiphytes are also woody. The undergrowth largely consists of seedlings, sapling trees, and young woody climbers.

The trees of the tropical forest are sometimes taller than those in temperate forests, but they do not reach the gigantic dimensions of North American redwoods or the largest eucalypts in Australia.

By way of further illustrating the contrast between forest types, in boreal and temperate forests, dominant trees frequently belong to just a few and sometimes a single species, but in tropical forests there is seldom

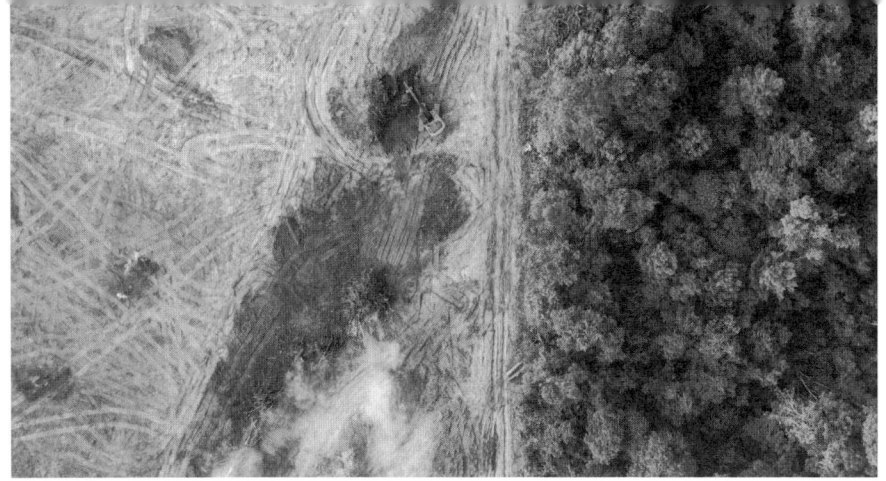

Aerial view of tropical forest being removed to make way for palm oil. About 150 million hectares of tropical forests have been destroyed between 1990 and 2018.

less than around 50 trees species present on any given hectare, sometimes well over a hundred. The richness and diversity of the tree flora is indeed an important characteristic of tropical forests.

Around half of the tropical forests that were present in 1800 have been cleared. Indeed, tropical forests continue to be cleared. As a result, some developing countries have high rates of greenhouse gas emissions. Papua New Guinea, for example, produces a third as much greenhouse gas as Australia, a country with four times the population and which still burns substantial quantities of coal to generate electricity.

The most recent consolidated data for global forest cover show a net loss of about 150 million hectares of forest between 1990 and 2018,[7] resulting in a one per cent reduction in forest land as a proportion of the global land area. However, the rate of annual net loss of forest has slowed from 0.18 per cent in the 1990s to 0.08 per cent in the period 2010–2015.

Globally 15 per cent of all human-caused greenhouse gas emissions result from the destruction of tropical forests.[8] If we could reverse this tragic trend, and by 2050 restore between 8 and 17 per cent of what we had removed, then between 40 billion and 200

Log harvesting. Only by continuing to grow a forest or a plantation and removing biomass from time to time in the form of wood can the forest continue to draw carbon indefinitely.

billion tonnes of carbon dioxide could be sequestered in this rehabilitated forest. This would make an impressive contribution towards balancing the carbon account book. What a great outcome that would be!

The role of a forest in storing carbon turns on the stage it is at in the growth cycle. When part of the biomass in a forest is destroyed or harvested, the forest takes in carbon to replace the carbon contained in the biomass that has been lost. After logs are harvested there is a period while the remaining and regenerating, or replanted, trees grow to occupy the space vacated by the trees that have been removed. There will be a period of modest increase in the quantity of carbon stored in a forest while the trees are regrowing. The quantity will plateau when the forest or plantation reaches its maximum biomass carrying capacity. At that time the amount of additional carbon being stored will be the same as the amount lost through tree mortality and decay. At that stage, the forest or plantation will be in an equilibrium state and the amount of carbon stored more or less constant. The essential point here is that only by continuing to grow a forest or a plantation and removing biomass from time to time in the form of wood can a forest continue to store carbon indefinitely.

Relying on trees for storage and related carbon accounting objectives runs into a road block when trees approach maturity and their growth slows and eventually stops. Well we know that trees do not grow on forever. A forest of mature trees locks up a large amounts of carbon, but because little new growth, if any, takes place in such a forest it doesn't absorb significant further carbon.

To store ever increasing amounts of carbon requires log harvesting, wood product manufacturing and tree regeneration or replanting. Even then the wood harvested needs to be stored long term—turned into durable timber products like houses, bridges or furniture.

It is an ecological reality that forests not used for wood production will transition from a period of growth when carbon is actively stored to a state when, as a mature ecosystem, growth and decay balance out. This was discussed in earlier chapters. Then

The use of biomass 'waste' to substitute for fossil fuels is gaining increased attention.

as trees begin to die and their crowns become less dense, the amount of decaying and dead matter increases and the forest may move from storing carbon to becoming a net producer of carbon dioxide. Typically, in mature forest ecosystems an equilibrium is reached when stored carbon dioxide and carbon dioxide production are pretty much in balance.

A significant point to make here is that natural forest systems tend to move towards a state of equilibrium, where tree growth is balanced out by decay, so the net result is that no additional carbon is stored. This state of equilibrium may be disturbed, for example, by fire—remember the 2019/2020 bushfires described in earlier chapters—where the dramatic loss of carbon dioxide from burnt trees tends to be followed by a period of growth and net carbon stored until a steady state of equilibrium returns once more.

As an example, it is estimated the forests across the European Union, together with the related wood products industry currently contribute to climate change mitigation at an amount equal to 13 per cent of total Europe Union country carbon emissions.[9, 10] This includes the forests' carbon sink function, as well as carbon stored in wood-based products, and the use of biomass "waste" to substitute for fossil fuels.

The amount of carbon that can be stored in a forest or plantation depends on the rate of growth and the species involved. Higher rates of storage are achieved by well-managed, fast growing species on good soils with adequate rainfall. Lower rates of carbon storage occur where rainfall is low, soils are poor and slow growing species are planted.

13

A LANGUAGE THAT THE STRANGERS DON'T KNOW

Trees have a completely different way of communicating

According to the dictionary, language is what people use to talk to each other. Looked at that way, humans are the only beings who use language, because the concept is limited to us. We all know that trees don't talk—so how then do they communicate with each other? They don't produce sounds, so unless you count branches creaking or leaves rustling there's nothing to hear.

But wait—it turns out that trees have a completely different way of communicating—they use scent. Scent as a means of communication is not totally unfamiliar to us—why else would we use deodorants and perfumes? Scientists believe pheromones in human sweat are a contributing factor when it comes to choosing partners. So, it seems fair to say that we possess an unspoken language of scent. Trees have demonstrated that they do as well.

Let's look at an example then. Giraffes on the African savannah feeding on umbrella thorn acacias move around quickly. In just mere

LEFT: Giraffes on the African savannah feeding on umbrella thorn acacia trees. Acacias start pumping toxic substances into their leaves to rid themselves of these large herbivores. RIGHT: Willow trees produce the defensive compound salicylic acid that is a precursor of aspirin. Tea made from willow bark relieves headaches and brings down fevers.

minutes the acacia starts pumping toxic substances into its leaves to rid itself of the large pesky herbivore. The giraffes get the message when the toxins arrive and move on to other trees. Not to trees close by though; they walk past a few trees and resume their meal some distance away.[1]

The reason for this behaviour is that the acacia tree being eaten gives off a warning gas—specifically, ethylene—to signal to neighbouring trees that a crisis is at hand. Immediately all forewarned trees start to pump toxins into their leaves. The giraffes have wised up to this game and move farther away to trees that are oblivious to what is going on, or else upwind where they can find acacias that have no idea that giraffes are looking for a meal.

The effectiveness of this kind of response depends on a capacity trees have for internal communication when threatened or hurt. Leaf tissue sends out electrical signals, just as human tissue does. However, leaf signals are not transmitted in milliseconds, like the human sensations which are communicated by nerves. Instead, the tree signal travels at the slow speed of around a centimetre a minute.[2] It may take an hour or so before defensive compounds reach the leaves to spoil a resident pest's or herbivore's meal.

A tree can send information from one part to another. If the roots find themselves in trouble, this information is transmitted throughout the tree, and can trigger the leaves to release scent compounds formulated to counter the threat at hand. Willow trees produce the defensive compound salicylic acid, which works in this way. Salicylic acid is a constituent of aspirin. Tea made from willow bark relieves headaches and brings down fevers.

Clearly trees communicate for similar reasons as human communities do—there are advantages in working together. On its own, a tree cannot establish a consistent local climate, and is at the mercy of wind and weather. But together many trees—a forest—create an ecosystem that moderates extremes of heat and cold, stores a quantity of water, and generates local humidity. It's the old story about strength in numbers.

What and how much information is exchanged between trees is a subject of ongoing research. For instance, the magazine *Nature*[3] reports that scientist Suzanne Simard and colleagues have discovered contact between trees of different species, even species in competition with each other. So, talk to a tree next time you are in a forest. It may answer you back, even if, to steal a line from Bing Crosby's Irish song *Galway Bay*, in "a language that the strangers do not know".

Trees are often connected to one another under the ground. Their roots may be in direct contact or it can be the fungal networks around tree root tips—which facilitate nutrient exchange between trees—are interconnected. Fungi while pursuing their own agendas work very much in favour of collaboration and equitable distribution of information and resources. The fungi wraps masses of tiny filaments called mycelia around a tree root; every rootlet of a tree may be completely encased in mycelia to the extent that the roots themselves won't even touch the soil.

The mycelia spread out like a web through the soil, connecting with the roots of other trees and forging a sophisticated underground network.

The benefits for the tree are enormous. Mycelia can get into places roots can't reach, increasing the area of absorption for the tree hundreds or thousands of times. The fungus seeks out nutrients and water to pass on to the tree, or to exchange with other trees. Working together, the fungi

Tree roots and fungus ... increasing the area of absorption hundreds or thousands of times.

and the tree can tap into sources of underground nutrients and water. In exchange for these essential services trees feed the fungi with sugar which keeps them alive. The fungi protect the roots from pathogens, filter out heavy metals, and enable information to be shared between trees.

Because tree defence mechanisms take time, a combined approach is crucially important for arboreal early-warning systems. The fungal connections transmit signals from one tree to the next, helping the trees to exchange news about herbivores, insects, drought, and other dangers.

To further illustrate the point let's look at an Australian example. Ribbon gums—*Eucalyptus viminalis*, also called manna gums or white gums— have been dying on a huge scale. In a sort of Bermuda Triangle in the Monaro region of New South Wales between Cooma, Nimmitabel and Jindabyne. Across more than 2,000 square kilometres every last ribbon gum is dead or dying, leaving a desolate landscape littered with skeletal remains of once healthy trees. The wave of destruction, first observed around 2005, has been quick and brutal.[4]

Dieback is not uncommon in ribbon gum forests, but this recent case is now recognised as one of the most significant and mysterious ever to hit Australia, with possibly worse still to come. In nearby Kosciuszko National Park the iconic snow gums—*Eucalyptus pauciflora*, also known as white sallee—are showing signs of the same decline.

The eucalyptus weevil has been devouring ribbon gum leaves.[5] Normally trees have a way of dealing with that kind of insect attack. If a tree loses its crown it will shoot out new epicormic foliage and these new leaves provide the tree with enough energy to start repopulating depleted mature leaves. However, if the insects keep attacking eventually the tree gets stuck in a lethal loop and runs out of energy stores needed to sprout new growth, succumbing and eventually starving to death.

By means of the network below ground we are discussing, trees feed neighbours and communicate dangers such as insect attack so that other trees can rally their defences. When these social networks are cut off trees lose this ability to communicate.

Eucalyptus weevil devouring ribbon gum leaves.

So, in ribbon gum forest, as elsewhere, trees need other trees around them—the more the merrier—for an efficient ecosystem. The canopies can protect each other from extreme wind, stop the heat of summer scorching the forest floor and help the whole community store more water and generate humidity creating its own local ideal habitat.

Existing ribbon gums have in the past been subjected to poor land management practices with clearing, over grazing and subsequent massive erosion stripping away the topsoil and compacting the ground. Rather than being absorbed by the soil, surface water forms into rivulets which become eroded gullies. Rising water tables have resulted in salinity levels soaring, and the application of superphosphate fertiliser has brought artificially high levels of nutrients to the soil, interfering with the trees' natural nutrient distribution systems.

Native grasses have been replaced with pasture grasses resulting in a loss of biodiversity. Then there are the impacts of climate change and drought. Once you realise the dead ribbon gums weren't always straggly lone individuals but part of an interwoven bushland, their apparent demise seems plausible, even inevitable.

Peter Marshall, a forest mycologist specialising in rehabilitating degraded land, says the forest underground is often overlooked. He has a particular interest in the symbiotic relationship between tree roots and a type of subterranean fungi mycorrhizae. "Without it our forests would be dead."[6]

Marshall suspects that a huge reduction in mycorrhizal fungi has had a drastic effect on the health of trees everywhere. "Once you know where to look you see it everywhere. The system has evolved over 100 million years and is utterly crucial to the success of Australia's landscape."

View across the Australian alps with dead snow gums in the foreground, Mt Hotham, Victoria.

Matthew Brookhouse led a research team surveying the extent of snow gum forest decline in the alpine regions of southern New South Wales—the true extent of which remains unknown. Dr Brookhouse suspects shorter winters could be playing a part in drying out the trees. "It's devasting. These trees are symbolic of a remote and wild place. They're an emblem for wilderness and we're watching them die."[7]

Professor James Trappe of Oregon State University agrees

that the literature on subterranean ecosystems is too often ignored. "We know that if there's damage to the soil and the mycorrhizas it sets the tree up for damage from insects. If trees are under stress—whether it be drought or insect attack or whatever—their ways of overcoming that stress depend on having a good root system."[8]

Many internal processes of plants and animals are regulated by chemical messengers. Roots absorb substances and bring them into the tree. In the other direction, they deliver the products of photosynthesis to the tree's fungal partners and send warning signals to neighbouring trees. But do trees have brains? For there to be something we would recognise as a brain, neurological processes must be involved and for these, in addition to chemical messages, you need electrical impulses. And yes—there are impulses that we can measure in a tree. So, can trees think, and are they intelligent?

Right now, most plant researchers are sceptical about whether such behaviour points to a repository for intelligence, the faculty of memory, and emotions, and if that exists it threatens to blur the boundary between plants and animals.

And so what? What would be so awful about that? The distinction between plant and animal is, after all, arbitrary and depends largely on the way an organism feeds itself: green plants photosynthesise and animals eats other living beings. The only other big difference is in the amount of time it takes to process information and translate it into action.

Mystical ancient laurel tree, Laurisilva Forest, Madeira Island, Portugal. Do trees have a mind? Are they intelligent?

Araucaria araucana trees, Patagonia, Argentina. A distinguishing feature of many South American landscapes, commonly called monkey puzzles. Because of their longevity they are often described as living fossils.

So trees are in low gear and animals much speedier. Why do we conclude that beings living life in the slow lane are necessarily incapable of the thought and feelings we attribute to the ones on the fast track?

We know that trees exhibit substantial tolerance to variations in climate. In the Southern Hemisphere it is generally accepted that the Araucariaceae family of conifers is one of the oldest, if not the oldest of conifer groupings. The genuses of the group *Araucaria* and *Agathis* are remnants of a group of conifers that once flourished north of the Equator as well. These trees retreated from their former dominant position in the world's flora as the climate fluctuated. Today they only have a southern distribution.

The Agathis genus now consists of 13 known species, usually called kauri after the Maori name for the New Zealand species. Agathis prefer warmer areas, although some exist as far south as the warm temperature moist evergreen forest of northern New Zealand. The main range of the now remnant population is the tropical moist evergreen forests of the southwest Pacific: the Malay Peninsula, the Malay Archipelago, the Bismarck, Santa Cruz, New Hebrides and Fiji archipelagos, New Caledonia and Queensland.

A distinguishing feature of many South American landscapes are *Araucaria araucana* trees, commonly called monkey puzzles. An evergreen native to central and southern Chile and western Argentina, they are the hardiest species in the genus *Araucaria* and because of their longevity are often described as living fossils. Their natural habitat is the lower slopes of the Chilean and Argentinean south-central Andes Mountains, typically above 1,000 metres. Juvenile trees exhibit a broadly pyramidal or conical habit which develops into the distinctive umbrella form as trees mature.

The genus *Araucaria* has a long and noble ancestry, and still consists of 25 species, all in the southern hemisphere. Palaeobotanists have traced the ancestors of the genus back to prehistoric times. Common to all species is the regimented geometry associated with the time of the dinosaurs. The trees radiate whorls of branches stiffer and spikier than those of any other conifers. Four species (*A. excelsa, A. bidwillii, A. columnaris* and *A. cunninghamii*) originate from the southern Pacific, including the rainforests of Queensland and Norfolk Island.

North of the Equator, European beech still grows from Sicily to southern Sweden. These regions have little in common, but birches, pines, and oaks have also demonstrated great flexibility, and when temperatures and rainfall fluctuate, along with many animals and plants they move from south to north and vice versa.

If climatic conditions change, the first trees to die out will be those that have the hardest time dealing with the new status quo. A few old trees will die, but most of the rest of the forest will remain. If conditions become more extreme, one tree species could even be wiped out without this being the end of the forest. Usually, sufficient trees remain to produce enough fruit and shade for the next generation even if the forest tree species composition changes.

An olive tree, perhaps thousands of years old. Trees live their daily lives at an incredibly slow pace.

Thanks to the migration of trees, the forest is constantly changing, but in human life spans this may not be readily apparent. An illusion of permanence is almost perfect in the forest, because trees are among the longest lived and slowest-moving organisms with which we share our world, and changes are observable only over the course of many, many human generations. Trees live their daily lives at an incredibly slow pace. It's no wonder that even though every schoolchild knows trees are living beings, trees are also categorised as objects.

It's okay to use wood as long as trees are allowed to live in a way that is appropriate to their species. That means that they should be allowed to fulfil their social needs, to grow in a true forest environment on undisturbed ground, and to pass their knowledge on to the next generation. Certainly at least some of them should be allowed to grow old with dignity and finally die a natural death.

We have learned that parent trees recognise and communicate with their kin, shaping future generations. In addition, injured trees pass their legacies on to their neighbours, affecting gene regulation, defence chemistry and resilience in the entire forest community. These discoveries have transformed our understanding of trees which were once regarded merely as competitive individuals and are now recognised as members of a connected and communicating community. There is a burst of scientific research directed worldwide at uncovering all manner of ways trees communicate with each other above and below ground.

PUTTING A PRICE ON NATURE

Impossible, but part of economic decision-making

The next couple of chapters deal with the intersection of economic theory and practice with the value of nature and climate change-related matters.

The French economist and anthropologist Jacques Weber expressed the perplexity experienced by economists faced with the problem of valuing nature or "natural capital". He said: "Whatever the price given to the Mona Lisa, that would say nothing about its value. The comparison in attempting to value nature with a work of art, which is unique and irreproducible, is very relevant."[1]

Thousands of copies of the Mona Lisa are sold every year in the Louvre Museum shop and elsewhere, at prices that vary according to the quality of the reproduction, its size, frame, packaging and so on. These copies are just like any other merchandise, but the Mona Lisa itself has no price and like the Notre Dame Cathedral, the works of William Shakespeare, and nature its real value can't be estimated. As well as being non-reproducible, nature is what is called a public good. That is, a good held in common for everyone. It is also indivisible. No more than cutting up the Mona Lisa into small pieces, like carpet samples, and selling them, might be contemplated slicing up nature and trading its components in a market. However, at least in economics and public policy this is where the commodifying of nature occurs.

While it makes no sense to put a price on nature, there are nevertheless many situations in which society must evaluate the costs of its use, or its destruction. Introducing the idea of an environmental good into decision-making should rightly reflect the use made of nature. In this sense, nature can indeed be valued and should be taken into account in economic decision making.[2]

When the oil tanker *Exxon Valdez* went aground in Prince William Sound, Alaska, it spilled 42,000 tons of crude oil into the Arctic Ocean producing an oil slick that seriously damaged the very fragile ecosystem in the region. Five years after the disaster Exxon Mobil was fined $US5 billion. After an appeal in 2008, the United States of America Supreme Court reduced the amount payable for damages and interest to $500 million. In the end, Exxon Mobil spent nearly $3.4 billion cleaning up the coast and wildlife as best they could, and on compensation for damage to fisheries.

The *Exxon Valdez* example makes it clear that we are far from knowing how to price environmental damage because there is no apparent consensus on the methods used for assigning a value to nature and establishing a cost for its use and value for its protection.

Exxon Valdez oil tanker aground in Prince William Sound, Alaska spilt 42,000 tons of crude oil into the Arctic Ocean producing an oil slick that damaged the very fragile ecosystem.
Source: AP Photos

When applying economic theory, price plays a central part. It reflects use values and enables assessment of the gains or losses in welfare that consumption or production of a good brings to an individual or to society. Let us look at this in a different way. If, for example, I decide to buy a traditional baguette (baked without any additives, unlike an ordinary baguette), at $1.15 instead of an ordinary baguette at $0.90, the additional pleasure derived from eating the traditional baguette is worth at least $0.25. If it is worth more, I realise a surplus and if it is worth less, I go back to the ordinary baguette.

Soca river and gorge, Bovec, Slovenia, popular rafting and kayaking destination. Such free environmental goods can be over used, or even destroyed.

The baker sets the price of the baguette on the basis of his costs, market conditions and what he believes I, or someone else is willing to pay. Price systems are ultimately based on these sorts of economic calculations.

Economic theory provides two fundamental tenets of welfare to calculate under what conditions society can maximise its collective welfare. This so-called theory of "general equilibrium" was formalised by the Nobel Economics Prize winners Kenneth Arrow and Gerard Debreu.[3]

The first welfare theorem asserts that if all parties behave competitively in markets where there is perfect information, then market equilibrium leads to a situation in which it is impossible to improve the situation of one or more party without worsening at least one other party's situation. This is known as the Pareto Optimum, named after the economist, Vilfredo Pareto.[4] The Pareto Optimum is therefore a criterion of efficiency in the use of resources in the economy. The conditions for achieving such efficiency in the allocation of resources are highly specific and restrictive, making it extremely difficult to transpose the theorem into real life. Information must be perfect. That is, we know everything about everything; people do not engage in strategic behaviour, and there is perfect competition. Also, all goods have a price—markets are all encompassing, and there are no public goods, or put another way, there is an absence of collectively-owned goods.

Nevertheless, the theorem does highlight one important element. People's decisions are based on a trade off or choice among different values reflected by prices. The more the various values are known the better the decision making. By increasing the capacity to measure the value of resources, there is a gain in economic efficiency if the markets function properly. But what happens when an environmental good remains unpriced or impossible to price?

The answer is found in the title of the influential paper *The Tragedy of the Commons* by Garrett Hardin published in 1968.[5] Hardin outlines how an environmental good being free leads to its overuse, or even destruction, in a society dominated by price-based market exchange. He draws on the example of a freely accessible communal pasture that, in a feudal village society, serves as social security.

So, every villager of whatever status, is free to take livestock to graze in the pasture. Though effective in an unchanging economy, this type of organisation does not support growth. Increasing the number of animals leads to overgrazing of the communal pasture. Because its degradation does not impose costs on any of the villagers individually, it is in each villager's interest to use the pasture for as long as any grass remains. Inevitably this leads to the destruction of the communal pasture or environmental good. Hardin reminds us that the issue of measuring the value of environmental goods is not just limited to cases of compensation, as in the case of the *Exxon Valdez*. In the absence of an explicit and agreed valuation, there is a great risk of over-exploiting nature, or destroying it. Why pay attention to it if its value is zero?

In a situation where the value of an environmental good is zero nobody bears the cost, and nobody assesses the consequences of any detrimental environmental consequences. Because no person or entity is debited this cost, it is not assigned a price in accounts or in consumption and production decisions. Nor does it normally influence behaviour. One of the challenges of environmental goods is that generally they are not commodities, and are difficult, if not impossible, to price and therefore remain unpriced. It may be regrettable that a monetary value has to be put on something for it to be taken into account, but until such time as the economy is regulated by mechanisms other than the market, an answer must be found if the tragedy of the destruction of environmental goods is to be valued, priced and if possible avoided.

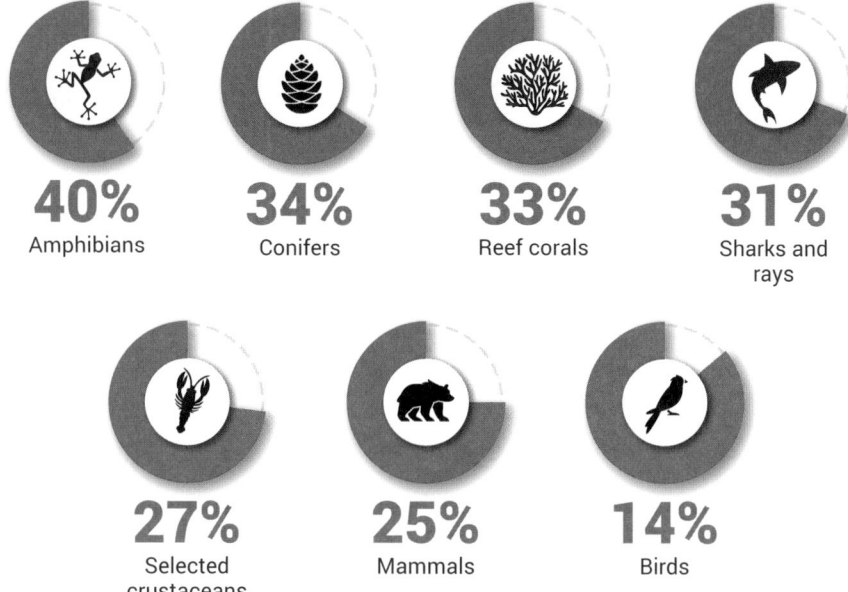

40% Amphibians   34% Conifers   33% Reef corals   31% Sharks and rays

27% Selected crustaceans   25% Mammals   14% Birds

One in four species is at risk of extinction (species assessed by the International Union for Conservation of Nature Red List). Source: IUCN Red List of Threated Species and BBC

How to solve the difficulty of pricing environmental goods has long bedevilled economists and public policy makers. Whether it is a matter of a natural resource, carbon pollution, or more generally an externality,[6] it is not easy to assess the different possible values of an environmental good. A forest may be used for the production of wood, but also for walking in today, tomorrow, or in ten years' time. In addition, a forest produces and reproduces biodiversity and plays a role in water regulation and in the carbon cycle. A lake can be used for power generation, recreation, irrigation, biodiversity and so on. Many other examples are available.

In the context of placing a value on the natural world, in his 2020 book, *A Life on Our Planet: My Witness Statement and a Vision for the Future*, David Attenborough[7] said:

Brown bear and cubs in North American forest. Pricing environmental goods has long bedevilled economists and public policy makers.

The Earth's last forests, rainforests, wetlands, grasslands and woodlands are, in fact, priceless. They are the carbon stores that we cannot afford to unlock. They offer environmental services that we cannot do without. They are home to biodiversity that we must not lose. How can we come to represent all that in our value systems?

Economists generally consider that an environmental good has a total economic value that can be broken down into three components:

1. A direct use value (the wood taken from a forest, or the welfare obtained from walking in it).
2. A non-use or indirect use value (the wood left for future generations, or the reserve of biodiversity that contributes to the collective welfare and heritage).
3. An intrinsic value that exists independently of human society (the value of the species living there and the ecosystem they support and that supports them).

Each of these components is linked to philosophical or spiritual foundations. As we have already discussed, nature plays a prominent role in terms of the worshipping and the ritualisation of prayers.

SELLING CARBON IN THE MARKET

Good idea? Some governments don't think so

When carbon dioxide is emitted into the atmosphere it creates a cost to the economy by accelerating climate change. The air pollution from burning fossil fuels also comes at a cost, since the smoke can contribute to cancer and other insidious diseases. In theory these costs can be estimated. For example, one recent study estimated the health effects of burning fossil fuels in the United States of America at about $US900 billion a year. Including those costs in the price of electricity would at least double the price per kilowatt-hour.[1]

Of course, in most of the world, polluting with carbon and other emissions has been free, with no consequences to the person or entity doing the polluting. The consequences are instead shared by everyone, and by future generations.

A clear solution to this issue, but one not generally embraced by governments, is some form of governance that makes polluters pay for the costs of their actions, rather than passing them along to society. A carbon price could solve this problem. Charging for the externality of the costs of carbon pollution would provide an incentive to pollute less.

Charging a fee for carbon emissions would shift incentives across the whole economy. It would result in saving money by using energy more efficiently. Also, large scale power sources that do not pollute, including nuclear power, would gain a competitive advantage compared with fossil

Coal-fired power station. When carbon dioxide is emitted into the atmosphere it creates a cost to the economy by accelerating climate change.

fuel energy sources which would carry their full costs via a price on carbon.

Carbon pricing is an efficient mechanism that would allow market forces to do the work, rather than having a government or private sector mechanism dictating how much energy individuals and businesses can use. A carbon price provides an incentive by charging fossil fuel-based energy providers for putting carbon dioxide into the atmosphere.

A tricky part of implementing a carbon price is setting the right price. William Nordhaus, winner of the 2018 Nobel Prize in Economics, researched the question of what price to charge for carbon pollution, given that too high a price leaves future generations with less money to fight climate effects in their own way, and too low a price leaves them with climate effects that will be more expensive to fight.

To simplify, Nordhaus estimates a low price of $US11 per ton of carbon dioxide worldwide would result in global warming of around 3.5° by the end of the century. A price closer to $US50 per ton would more likely meet the 2° maximum temperature increase target, if everyone in the world participated.[2]

Many companies have now begun building a carbon price into their internal planning assumption as they contemplate long-term investments. A survey of 21 electric utilities in the United States in 2012 found that 16 assumed a future carbon price, averaging almost $US25 per ton for 2020. One report[3] gives a spectrum of internal prices applied to operations in companies such as Microsoft ($US6–7 per ton); Disney ($US10–20); Google ($US14) and Exxon Mobil ($US60–$US80), among others.

Collecting a carbon price via a tax would not be too difficult, because it could be charged at the first point where the fossil fuel enters an economy. Say at the well-head where fossil fuels are produced or the port the fuel was imported through. Instead of charging each car driver for the carbon dioxide that comes out of their exhaust pipe, the government could charge a fee on the crude oil before it is even turned into a fuel.

To gain support for a carbon tax, some advocates favour using the revenue to reduce other taxes, or send rebate cheques to citizens. This revenue-neutral concept goes over well with those concerned that a carbon tax will be just another way to raise revenue and expand the scope of government.

The Canadian province of British Columbia has successfully implemented a revenue-neutral carbon tax.[4] The tax began in 2008 at about Can$9 per ton of carbon dioxide and rose to about Can$27 per ton by 2012. The tax was revenue neutral because other taxes were cut by the same amount that the carbon tax brought in. Corporate taxes dropped from 12% to 10%. Emissions fell modestly and the province's economy performed well, with most businesses coming to support the tax.

On the whole, carbon pricing has great potential to help slow climate change. It takes effect quickly and changes behaviour across the whole economy without any need to tinker with each wind turbine or electric car charging station in an effort to engineer similar outcomes. Introducing a carbon price, or tax to assist in confronting climate change would seem like a no-brainer. It is a real pity therefore that many governments have been reluctant to wear the political acrimony of doing so.

Carbon tax signage. Carbon pricing has potential to help slow climate change.

Cargo plane taking off with turboprop engine smoke emission. Emission allowances can be traded amongst companies to match their pollution levels.

There are alternatives to selling and implementing a carbon price or tax. An alternative method to price carbon is "cap and trade". Governments set a limit on how much carbon can be released—the cap—and issue permits allowing companies to do so. The permits might be free, or sometimes they are auctioned off by governments. Companies can then trade the permits, creating a market for carbon dioxide pollution that encourages efficiencies to be achieved.

The European Union has such an emissions trading scheme. This scheme grants carbon emission allowances to 11,000 industrial and power installations as well to airlines. These carbon emissions allowances cover almost half of the European Union's carbon emissions. The European Union's emissions trading scheme aims to cut European carbon dioxide emissions by 20% compared with 1990 levels by 2020, and 40% by 2030. During the 2013–2020 period, emissions caps have been reduced by a bit less than two per cent each year. The aviation sector has separate less ambitious targets.

Allowances are initially allocated or sold to carbon polluters by governments. Companies can then trade them amongst each other as needed to match their pollution levels. A company with too high a level of emissions can either become more efficient, that is emit less carbon, or buy allowances from a less polluting company.

Of course, there are some loopholes. For example, a company might just shift production to another part of the world not covered by carbon pricing. This is an ongoing concern to the European Union.

One advantage of cap and trade schemes is that a government authority can set the actual allowable level of carbon pollution, whereas under a carbon price mechanism, nobody can say for sure how many people will pay a given price and therefore how much carbon dioxide will really be emitted into the atmosphere.[5] But as emission trading systems, and complementary efforts indicate, it is hard for governments to assess how many pollution permits to allow and how much to initially charge for them.

A carbon price is simpler than an emissions trading system and affects everything across the entire economy. It's easy to administer, and acts quickly. As it restructures the economy away from fossil fuels, it can boost economic growth and create jobs, and is centred around renewable energy.

THAWING UP IN THE ARCTIC

The Arctic is warming more than twice as fast as the rest of the planet

Unfortunately, this chapter is a continuation of the scary climate change theme. We will start up in the Arctic, then crunch some more disturbing figures.

The Earth's five warmest years since the late 9th century have occurred since 2014. Troublingly the Arctic is warming more than twice as fast as the rest of the planet as it loses the sea ice that keeps it chilled.[1] Successive hottest years on record in the Artic have been confirmed by satellite monitoring showing that sea ice had shrunk to the lowest extent ever witnessed.

Today the Arctic landscape is greener with fewer caribou and reindeer, with warmer summers and more mosquitoes. However, the most visible and disturbing change has happened at sea, with summer sea ice—the floating expanse that still covers much of the Arctic Ocean during the region's brief thaw—disappearing at an astonishing rate.

While floating sea ice always shrinks during warmer months and expands again with the return of the cold, the scale of the summer ice loss has been unprecedented. Researchers believe that this process is actually speeding up. NASA scientists estimate that on average the Arctic ice

LEFT: Melting Arctic glacier in Svalbard, between mainland Norway and the North Pole. Summer sea ice—disappearing at an astonishing rate. RIGHT: Close-up of permafrost melting. As Arctic soils soften and slump, they release quantities of carbon that have been locked up in frozen soil for thousands of years.

shrinkage is about 50,000 square kilometres of ice a year and predictions are that it will continue much faster than anyone thought.[2] Scientists say the Arctic Ocean is very likely to be ice free in the summer before 2050.

"It's the largest Arctic ice loss in human history and it was not predicted by even the most aggressive climate models," said Jonathan Markowitz, a professor of international relations at the University of Southern California.[3]

The sea ice cover of the Arctic Ocean has a major influence on the broader climate system and on regional warming. As the area of summer ice melt expands, darker ocean water is exposed, absorbing rather than reflecting incoming sunlight and warming the region even further. For this reason, a rapid decline in sea ice extent is of major concern for both the Arctic and for the global climate system.

Research suggests that carbon stored in the Arctic will escape faster as the planet warms. The unexpected speed of Arctic warming, and the movement of meltwater through the polar landscape, is leading to the likelihood that for every one-degree rise in the Earth's average temperature, permafrost may release the equivalent of four to six years' worth of emissions from coal, oil and natural gas. This is double to treble what scientists thought a few years ago.[4] Another worrying aspect is that as Arctic soils soften and slump, they're releasing quantities of carbon that have been locked up in frozen soil for thousands of years, accelerating the rate of climate change.

Permafrost is ground that normally remains frozen all year. Permafrost is capped by a couple of metres of soil and plant matter. This soil layer normally thaws each summer and refreezes in winter and so protects the underlying permafrost from rising atmospheric temperatures.

Globally permafrost stores up to 1600 gigatons of carbon, nearly twice the amount stored in the atmosphere.[5] Until recently research presumed permafrost would lose at most 10 per cent of its carbon, and even that amount, it was thought, could take as much as 80 years to be lost. However, when the soil surface layer stops freezing in the winter, things speed up, enabling microbes to feed on thawed organic material, and emit carbon dioxide year-round, instead of just for a few short months each summer. Within a few decades permafrost could be as big a source of greenhouse gases as China, the world's largest emitter of carbon dioxide in 2019,[6] a threat that has yet to be fully accounted for in climate models.

Scientists are now discovering landscapes with permafrost that once thawed a few centimetres a year are now rapidly thawing down to about three metres, creating wetlands and greatly accelerating greenhouse gas emissions.

If permafrost continues to thaw, deeply buried ice that forms crystalline "cages" around methane molecules will also start to melt. An estimated 400 billion tons of these frozen methane deposits, known as clathrates, lie a few hundred metres beneath the ground, and even more are found beneath the world's oceans. All this deep-frozen natural gas, estimated at least to be equal to known conventional gas and oil reserves would,

Ice cap melting, near Kangerlussuaq, Greenland. Ice melting is accelerating and climate scientists have repeatedly said that all signs point to the need for urgent and audacious action.

Coral colony bleached by warming water temperatures. Repeating bleaching can cause the death of coral and the conversion of reefs to an algae-dominated ecosystem.

once the ice that encases these reserves starts to melt, release quantities of methane that might ratchet up global warming to levels not yet predicted.

The human community has spent decades ignoring the evidence of climate change, not responding to the warming Arctic, and hoping that things won't get too bad. This in spite of the fact that climate scientists and permafrost experts have repeatedly said that all signs point to the need for urgent and audacious action.

To combat the effects of climate change in a number of developing countries crops will need to be developed that need less water, and can tolerate higher temperatures. For already drought-prone countries like Australia, technologies to use water more frugally, along with bushfire mitigation and suppression will need scientific effort and resourcing. Housing and building construction will need to place increased emphasis on storm resistance and energy efficiency.

Extreme weather events, the decay of Arctic ice, and the acidification of the oceans caused by the absorption of carbon dioxide are all proceeding at unanticipated rates. Indeed, the situation is now so severe that a report from the Intergovernmental Panel on Climate Change[7] concludes that global average surface temperature is unlikely to drop in the first thousand years after greenhouse gas emissions are cut to zero.

Some of this is bleak for sure. However, it is well past the time to start being an alarmist. At home here in Australia record warm sea surface temperatures and widespread coral bleaching are threatening the survival of the Great Barrier Reef.[8] Prolonged hot temperatures have also contributed to major algae blooms and exceptionally low flows in inland river systems, and hot and dry conditions have been a significant factor in unprecedented bushfire destruction.

Australia, and the rest of the world, are in uncharted territory when it comes to the warming of the world's climate. Science-based studies confirm that adverse climate change is becoming an environmental and economic reality right now, reinforcing the urgency of calls for more immediate action to confront the here-and-now impacts on the everyday lives of the planet's inhabitants.

The price of inaction or ineffective action will obviously be much higher because of the damage and human suffering that unconstrained climate change will unquestionably cause. Even if decisive action to cut greenhouse gas emissions is initiated many of the changes that are already underway simply cannot be halted. So, like it or not, we need to learn to live, if we can, with the consequences of climate change. This means determining the probable effects of climate change and taking action to minimise its more severe impacts.

However, there is a lack of urgency and willingness, even in the face of the overwhelming evidence, to start to get serious. The second largest emitter of greenhouse gases, the United States of America, has until 2021 been the least willing to do anything about it. Even if we take strong action today, temperatures will continue to rise in the immediate future. Little action means that temperatures will increase even quicker. Action to reduce emissions sounds good politically, but is I fear too little, too late.

NATURE RELIEVED OF ITS HUMAN BURDEN

How nature might respond if granted half a chance

Approaching the serious part of the book, we need to confront the brutal reality that humanity, as we currently know it, is more likely than not to vanish from the face of the planet. The curiously named poem Drinking Song of the Sorrow of the Earth, by Chinese poet Li Tai Po (701 to 762 AD) seems to have been well ahead of its time. In part it says:

The heavens are forever blue and the earth
Will stand firm for a long time and bloom in spring.
But you, Man, how long will you live then?

So now let's flick the switch. Just how would the rest of nature respond if it were suddenly relieved of the relentless burden humanity heaps on the planet and our fellow organisms? How soon could the climate return to where it was before the Industrial Revolution and we pumped billions of tons of carbon dioxide skywards?

Could nature ever obliterate all our traces? How would it undo our massive, seemingly indestructible cities and infrastructure, and reduce our myriad plastics and toxic synthetics back to benign, basic elements? Or are some so unnatural that they are indestructible and will endure for millions of years?

Times Square, New York. Could nature ever obliterate all trace of us? How would it undo our massive, seemingly indestructible cities?

Perhaps it might not really be such a big task for Mother Nature. As outlined in earlier chapters, she has been through serious losses before, and refilled empty niches. While the human impact on the planet has been monumental, there are still places that remain essentially as they would have been if we were not here. Inevitably they invite us to wonder how nature might respond if granted half a chance.

So, let's think about this for a moment. Looking at the moist temperate climate of the Northern Hemisphere, birch and aspen trees are known to invade fallow fields quickly. Farmland can give way to woodland within just a couple of decades. Then under the canopy of the pioneering trees, oak, maple, linden, elm, and spruce trees will regenerate. Given say 500 years without people, a dense forest could return.

The Chernobyl disaster was a nuclear accident that occurred on 26 April 1986 at the Chernobyl Nuclear Power Plant, near the city of Pripyat in the north of Ukraine. In terms of cost and casualties, it is considered the worst nuclear disaster in history.[1]

Within a year of that disaster, the number of long-term evacuees passed 135,000. The years between 1986 and 2000 saw the near tripling to approximately 350,000 of the number of persons permanently resettled from the most severely contaminated areas.[2] After the disaster, four square kilometres of pine forest downwind of the reactor turned reddish-brown and died, earning the name of the "Red Forest". Some animals in the worst-hit areas also died or stopped reproducing. A significant economic impact at the time was the removal of 784,000 hectares of agricultural land and 694,000 hectares of forest from production.

More than 30 years have passed since the Chernobyl meltdown. Hundreds of billions of dollars have been spent on clean-up and literally untold thousands of people have died, been injured, or are sick.

An area originally extending 30 kilometres in all directions from the nuclear power plant is officially called the "Zone of Alienation". It is now uninhabited, except by about 300 residents who refused to leave or have crept back to their homes. The area has largely reverted to forest, and been overrun by wildlife because of a lack of competition with humans for space and resources.

Another Russian radioactive contamination accident that predated Chernobyl by three decades was the Kyshtym disaster on 29 September 1957 at Mayak. This was a plutonium production site for nuclear weapons, located at the city of Chelyabinsk-40 (now Ozyorsk) in Chelyabinsk Oblast. Measured by radioactivity released the disaster is second only to Chernobyl. At least 22 villages were exposed to radiation from the Kyshtym disaster, and around 10,000 people had to be evacuated. The incident occurred when the cooling system on a tank containing about 80 tons of liquid radioactive waste failed. The temperature rose resulting in evaporation and a chemical explosion of the waste, consisting mainly of ammonium nitrate, which blew the lid off the 160 ton concrete tank. Most contamination settled near the site, but a plume of highly radioactive particles extended over hundreds of kilometres.[3] The fallout of nuclear material spread to more than 52,000 square kilometres, occupied by at least 270,000 people.

In 1968, the East Ural Nature Reserve was established adjacent to the site, covering 16,600 hectares of the West Siberian Plain, from the Ural Mountains to the Central Siberian Plateau. Where villages and farmland were vacated is now a region of extensive boreal conifer forest and wetlands with diverse wildlife. The reserve serves to buffer the contaminated area, and for the long-term scientific study of the effects of radiation on forest ecology on the east slopes of the southern Ural Mountains.

Perhaps surprisingly, despite high radiation levels in the area today, wildlife populations around Chernobyl are growing freely in the absence of humanity. While there seems to be disagreement about the extent to

LEFT: Chernobyl nuclear power plant, Reactor 4, with containment sarcophagus. Both in terms of cost and casualties, it is considered the worst civil nuclear disaster in history. RIGHT: Pripyat city three decades after the nuclear meltdown. Apartment complexes and other buildings are starting to disappear beneath a green tree canopy.

which any populations can weather the radiation in the long run, for now, the animals are thriving, and parts of the exclusion zone have become a haven for biodiversity.

Taking a walk around the city of Pripyat today, three decades after the nuclear meltdown, if you want to risk it, you will immediately be struck by the advancing forest. Streets and open spaces are being occupied by trees. Apartment complexes and other buildings are starting to disappear beneath a green tree canopy.

Researchers have identified black bears, lynxes, European bison, boar and Przewalski's horses in growing numbers.[4] An unexpected side effect of evacuating people from the area has been to create a wildlife sanctuary. Is it that some animals, and plants, may be more resilient to radiation than thought? Or that the effects of the world's worst nuclear disaster are less damaging to the natural world than the continuing presence of people? Is this a lesson of hope, or a warning about our everyday impact on the planet?

Another interesting case with some similarities is the Demilitarized Zone (DMZ) separating the two Koreas. Supported by its Chinese and Soviet communist mentors, North Korea invaded the South in 1950. Eventually, United Nations forces pushed it back, and in 1953 a truce ended what had become a stalemate along the original dividing line, the 38th parallel. A strip two kilometres on either side of this dividing line became the no-man's-land, known as the DMZ.

LEFT: South Korean soldiers stand guard at the DMZ, the 38th parallel dividing line. A strip two kilometres on either side of this dividing line became no man's land. RIGHT: Asiatic black bears. Gradually the DMZ has been occupied by wildlife that had practically nowhere else to go.

Much of the DMZ runs through mountains taking in rivers and streams where, for thousands of years before hostilities began, people lived, built homes and farmed. Their abandoned paddies are now thickly sown with land mines. Since the armistice in 1953, other than occasional military patrols tippy-toeing through the minefields, or desperate, fleeing North Koreans, humans have barely set foot there.

In the absence of humans, the DMZ has been gradually occupied by wildlife that had practically nowhere else to go. Though inadvertent, one of the world's most dangerous places became one of its most biologically important—refuges for wildlife that might otherwise have disappeared. Asiatic black bears, Eurasian lynx, musk deer, Chinese water deer, yellow-throated martens, an endangered mountain goat known as the goral, and the nearly vanished Amur leopard cling here to what may only be temporary life support.

So perhaps if everything further north and south of Korea's DMZ were also to become a world without humans, these animals might have a chance to spread, multiply and reclaim their former realm.

In Eastern coastal regions of Australia where eucalypt species can aggressively recolonise grassland joining forests, farmers in northeast New South Wales have run foul of native vegetation legislation after leaving paddocks unstocked for a couple of years only to find on their return to what was once grassland is now an emerging forest of young blackbutt, spotted gum, ironbark, tallowwood and other eucalypt saplings that can only be removed with government approval before the cows can come back in.

We know that for many millions of years sheets of ice have been shrinking back and spreading forth from the poles as the global climate changed over geological time. The reasons for these climate shifts are thought to have been continental drift, the Earth's mildly eccentric orbit around the Sun, its wobbly axis, and swings in atmospheric carbon dioxide levels. In the last few million years, with the continents basically where they are positioned today, ice ages have recurred fairly regularly and totalled upwards of 100,000 years, with intervening thaws averaging somewhere around 20,000 to 30,000 years each.

In the past 20,000 years, as countless species, including most large mammals, became extinct, mice, rats, shrews and other small fur-bearing creatures have mostly managed to survive, as did marine mammals. Throughout the world, large land mammals took an enormous, lethal deathblow, vanishing in a geologic twinkling of about a thousand years.[5]

TOP: A herd of woolly mammoths grazing in the early morning frost. Throughout the world, large land-based mammals took a lethal deathblow, vanishing over a period of about a thousand years. CENTRE: Doedicurus (left) and glyptodon. The glyptodonts resembled armour-plated Volkswagens, with tails that ended in spiked maces. ABOVE: A sabre-toothed tiger confronts an early human. Along with many other large mammals, the sabre-toothed tiger disappeared at about the same time as well armed, clever humans took up residence.

The reason is pretty simple—when people moved out of Africa and reached other parts of the world large unsuspecting mammals became the favourite item on their dinner menu.

Among the missing legion of animal heavyweights are the giant armadillos and their even-bigger, heavily armoured relatives, the glyptodonts, resembling armour-plated Volkswagens, with tails that ended in spiked maces. There were giant short-faced bears, nearly double the size of today's grizzly bears. Giant beavers, as big as today's black bears, the American lion that was considerably bigger and swifter than today's surviving African species. Likewise, the dire wolf, the largest of canines, with a massive set of fangs.

Probably the best-known amongst the ranks of extinct colossi, the northern woolly mammoth, was only one of many kinds of elephant-like animals, including the imperial mammoth, largest of all at ten tons, and the hairless Colombian mammoth which lived in warmer regions. Mammoths were grazing animals, evolved to steppes, grassland, and tundra. In addition, three genera of American horses have all gone. Multiple varieties of North American camels, tapirs, numerous antlered creatures ranging from dainty pronghorns to the stag moose, which looked like a cross between a moose and an elk, but was larger than both, along with the sabre-toothed tiger and the American cheetah. All gone and all pretty much at the same time as well armed, clever humans took up residence.

Perhaps that mass extinction process may have started in Australia about 50,000 years ago, as humans arrived. Our sketchy evidence comes from uncovering and interpreting puzzles locked in fossil remains of the time when the first footprints were made on the northern coastline. Scientists reckon there is fairly compelling evidence that the original Australians came from the north out of Asia and across land and water into Australia somewhere between 50,000 and 100,000 years ago.

With the monsoon winds blowing from the north west there was a fair chance that even a flimsy raft constructed using bamboo, launched from somewhere in the Indonesian island chain could have bobbed across to the coast of Australia. At times of lower sea levels Australia was larger than it is today, and the stretches of ocean that separate it from northern neighbours would have been smaller or in some cases non-existent.[6]

So, it seems more than likely that the first landings in northern Australia were made by small groups, perhaps a few families who once lived and traded along the islands to the northwest and gradually established themselves as the occupiers of coastal areas of the newly discovered Australian continent.

The environment that confronted those first Australians must have been truly breathtaking. The continent would have been teeming with life, and with the pulse of growth and abundance, from the vibrant plant and animal life of the tropical north through to the diverse coastal forests further south and more austere landscapes of the tablelands and interior. So, it is not surprising that these people decided to stay and establish communities. Theirs is a remarkable story—a continuity and a link with the human past that today are found nowhere else.

In Australia, as in other continents early humans encountered animals that had no reason to suspect that small, seemingly insignificant erect primates were particularly deadly. Too late, they learned otherwise. As we have already related, even our ancient ancestors had already been fashioning primitive weapons of mass destruction.

By the time a group of them arrived at the threshold of America 13,000 years ago they had been "modern humans" for at least 50,000 years. With enhanced intellect they had mastered not just the technology of attaching fluted stone points to wooden shafts, but also to use a spear-thrower lever or *atlatl* that enabled them to propel a spear fast and precisely enough to fell seriously large animals from a relatively safe distance. The Australian

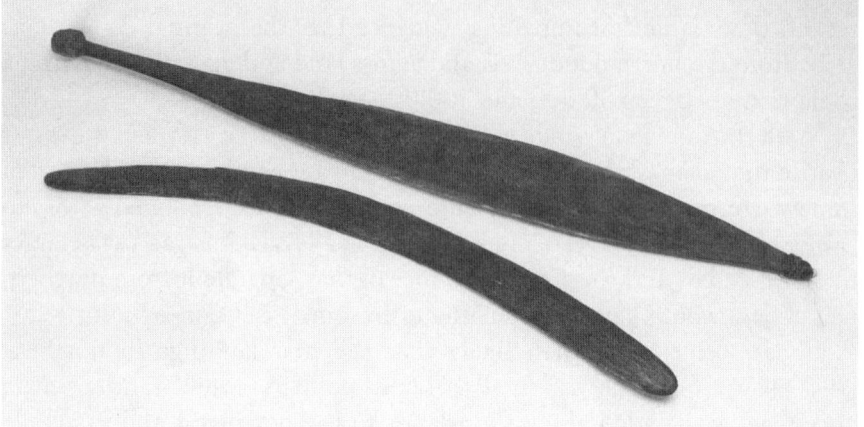

TOP: Early humans were particularly deadly, having fashioned primitive weapons of mass destruction. ABOVE: Woomera and boomerang. Aborigines devised ingenious weapons with lethal effect.

Aborigines had also fashioned the woomera, a spear-throwing device with similar lethal effect.

On another, far distant island, is a further hint that, had humans never evolved, now well and truly extinct large mammals might be around today. During the last ice age, Wrangel Island, a wedge of rocky tundra in the Arctic Ocean, was connected to Siberia. It was so far north that humans trekking across to Alaska completely missed it. As warming seas rose again Wrangel Island was isolated from the mainland, and its population of woolly mammoths became stranded and forced to adapt to the limited resources of an island environment.[7] During the span in which humans went from caves to building sophisticated civilisations Wrangel Island's mammoths lived on, a dwarf species that lasted 7,000 years longer than mammoths on any continent. They were still alive 4,000 years ago, when the Pharaohs ruled Egypt and built the pyramids.

By contrast, when the remarkable seagoing Polynesian navigators travelled into the Southern Ocean to reach New Zealand about 1300AD, they encountered moas—birds that stood over three metres tall—and at 250 kilograms weighed in twice as much as an African ostrich. The entire eleven species of moa were despatched within a couple of centuries by these Polynesians (the Maori) who colonised New Zealand, the last major landmass that humans discovered. By the time Europeans appeared in New Zealand some 350 years later, piles of big bird bones and Maori legends were all that remained of these giant, flightless, defenceless birds.[8]

Other massacred, flightless birds include the dodo of the Indian Ocean's Mauritius Island, famously wiped out within a hundred years of being clubbed to death and cooked by Portuguese sailors and Dutch settlers the bird never learned to fear. Because the penguin-like great auk's range stretched across the upper Northern Hemisphere, its extinction took longer, but hunters from Scandinavia to Canada still managed to exterminate them all.

Yet how is it possible that in less than a millennium early humans decimated America's rich megafauna? Surely Africa had even more people, and for a lot longer. If so, why does Africa still have its famous big-game menagerie? Why aren't Africa's big mammals now also extinct? The likely answer is that in Africa humans and large animals evolved together. Unlike the unsuspecting American, Australian, New Zealand and Caribbean herbivores who had no inkling of how dangerous our early ancestors were to their health when they unexpectedly arrived, African animals had the chance to adjust as human presence increased.

Animals growing up with human and other predators learn to be wary of them, and they evolve ways to elude them. With so many hungry neighbours, African animals learned that massing in large flocks made it harder for predators to isolate and catch a single animal. It also meant that some are available to look out for danger while others feed. A zebra's stripes help befuddle lions by getting lost in a crowded optical

Zebra herd. A zebra's stripes help befuddle lions by getting lost in a crowded optical illusion.

illusion. Zebras, wildebeest, and ostriches have forged an alliance on open savannas to combine the excellent ears of the first, the acute sense of smell of the second, and the sharp eyes of the third.

However, unfortunately today the balance that evolved between humans and wildlife in Africa has tipped out of control as there are now too many people and too many cows. Elephants and other large animal species have been squeezed into too few spaces making them easy targets for hunters and poachers.

Will baboons or chimpanzees evolve to assume the niche in the ecosystem if humans vanish? Has their cranial capacity lain suppressed because human ancestors got the jump on them, and were the first to climb out of the trees? With us no longer on the scene will the mental potential of baboons or chimpanzees surge and push them into an accelerated evolutionary scramble to become the dominant biped? The *Planet of the Apes* film series may not be all that far fetched.

Many insects will be major beneficiaries of human passing. Practices directed at the mass extermination of mosquitoes would stop. Humans were targeting mosquitoes long before the invention of pesticides, by spreading oil on the surfaces of ponds, estuaries, and puddles where they breed, to deny baby mosquitoes enough oxygen. Other methods of anti-mosquito chemical warfare included the aerial spraying of DDT,[9] only banned in some countries. With humans gone, billions of the flying insects that would otherwise have died prematurely would live. Among secondary beneficiaries would be many freshwater fish species, where mosquito eggs and larvae form a significant component of their food chains.

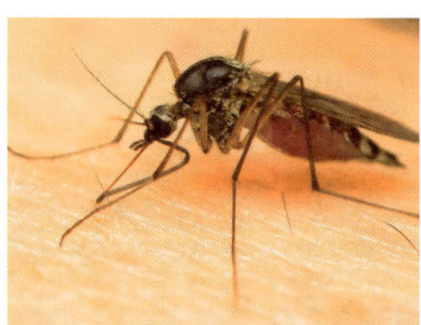

Macro-image of a mosquito on a human hand sucking blood. Humans were targetting mosquitoes long before the invention of pesticides, by spreading oil on the surfaces of ponds, estuaries, and puddles. Other methods of anti-mosquito chemical warfare include the aerial spraying of DDT.

With humans no longer on the scene oak shoots would sprout all across Europe from the acidic former fields of wheat, rye and barley within a few decades. Boars, hedgehogs, lynx, bison and beavers would spread, with wolves moving up from Romania and, if Europe cools, reindeer coming down from Norway. Crops, plants and livestock species would in all probability disappear within a century or two.

If humans were to go tomorrow, probably enough wild predators would remain to out-compete and demolish most of our domestic animals, though a few species that have proved impressively resilient in the wild indicate to us that there would be feral exceptions. Descendants of escaped wild horses, the burros of the American Great Basin and the brumbies of the Australian high country, occupy ecological niches that they may well retain. Dingoes, for so long Australia's top predator, can thrive in the wild, even though their ancestors were domesticated companions to early Southeast Asian traders. Perhaps some smaller domestic species of birds and mammals would survive. The world would mostly return to looking as it did before humanity came along—like a wilderness.

Of course, if everyone on Earth disappeared, more than 400 nuclear plants would overheat one by one. Some would burn, and the rest would melt. Spillage of radioactivity into the air, into rivers and the sea would be formidable, and enriched uranium would last far into future geological

Australian brumbies (wild horses). The brumbies of the high country occupy ecological niches that they may well retain.

Tihange Nuclear Power Station, Belgium. One by one nuclear plants would overheat. Some would burn, and the rest would melt. Either way, the spilling of radioactivity into the air, and into rivers and the sea would create a hazard for thousands of years.

time, a hazard for thousands of years to non-human victims who approached too closely.

In just decades with no new chlorine and bromine leaking skyward, the ozone layer would recover and ultraviolet levels subside. Within a few centuries, as most of our excess industrial carbon dioxide was locked away, the atmosphere and oceans would gradually cool. Heavy metals and toxins would be diluted and gradually flushed from the system. After plastic fibres and polychlorinated biphenyl (PCB, an organic chlorine compound once widely deployed in coolant fluids) are recycled a few thousand times, anything truly intractable would end up buried, to one day be metamorphosed or subsumed into the planet's mantle. Long before that every dam on Earth will have silted up and spilled over. Eventually our world would start over again.

To us still here on Earth, right now, the question more crucial than what will happen without us is whether we humans can make it through what many scientists call this planet's latest great extinction—make it through, and bring the rest of life with us rather than tear it down.

ENTER THE CORONA VIRUS

Nature is sending us a warning

In late 2019 from Wuhan in China the corona virus has travelled around the world. As vaccines started to be rolled out in the early months of 2021, a stocktake of the impact of the virus is chilling. As at about the middle of 2021 the number of people struck down by the virus was approaching 200 million, with four million dead and counting. In the context of this book, is there a connection between the corona virus and climate change?

George Monbiot[1] suggests that a bubble has finally been burst, but will we now attend to the other threats facing humanity? He wrote:

This could be the moment when we begin to see ourselves, once more, as governed by biology and physics, and dependent on a habitable planet. Never again should we listen to the liars and the deniers. Never again should we allow a comforting falsehood to trounce a painful truth. No longer can we afford to be dominated by those who put money ahead of life. This corona virus reminds us that we belong to the material world.

For years now influential private sector institutions have warned that climate change poses significant financial and environmental risks if we continue down the path of a business as usual carbon-based economy. In an opinion piece former politician John Hewson[2] wrote that the corona

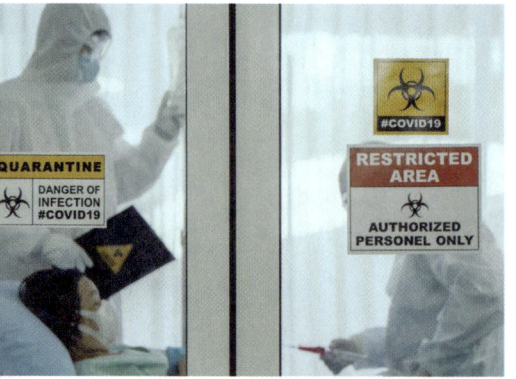 

LEFT: The corona virus has infected over 220 million people world wide with more than 4 million fatalities. RIGHT: Protective face masks in the workplace. Restraints enforced on personal liberties and national economies are directed at containing the corona virus infection.

virus pandemic should be seen as a dress rehearsal for what awaits if we continue to ignore the laws of science, the physical world and the demands of catastrophic threats, such as climate change.

He asserts that just as Australia was disturbingly unprepared for the recent bushfires and drought, even though fire and drought had recurred many times before, the globe was unprepared for the corona virus, even though there had been many warnings of such a risk. Hewson believes that governments and policy authorities seem incapable of accepting scientific and other evidence, and fail to listen to clear warnings and predictions. They are also generally unwilling to plan for how to avoid or manage a series of catastrophic risks that are mounting and threaten our living standards and lifestyle, and in the end, survival.

Previous chapters in this book have told us that humans have—even if largely unintentionally—threatened significant harm to ourselves and to the planet, by prioritising economic expansion and population growth, but largely ignoring their social, political and environmental consequences.

We know that fear is a great motivator, but who would have thought that in the space of just a few weeks the world could change as dramatically as it did in response to the corona virus. Politicians, businesses and communities have accepted dramatic changes to the norm that would have been unthinkable previously. We stopped travelling, we mostly stopped commuting to work. We adopted social distancing and stay-at-home behaviours.

Our political leaders threw ideology to the wind, and came up with policies and solutions that put people before politics. This has been an inspiring indication that the global community can embrace essential change. The risks emerging from the corona virus were varied, complex, global and catastrophic, and the solutions needed to be national, globally collaborative and multi-disciplinary.

Solutions to these risks depended, not just on government policy and corporate activity, but also on the daily actions of billions of people. Much of the virus-induced behaviour was motivated by the need to see civilisation not only survive the pandemic, but eventually prosper.

So, this means longer term, many existing systems and practices that have been taken for granted—our economic system, our food system, our energy system, our transport system, our production and waste systems, our governance mechanisms, our community lifestyles and our relationship with nature—must all undergo searching examination and reform if humanity is to survive.

Most would never have envisaged the restraints enforced on personal liberties and national economies directed at containing infection. Perhaps amazingly the world mostly accepted, and adjusted to these needs. To learn from this experience, we need to become proactive, anticipating and preparing to address global risks, and starting to develop the policy responses that will maximise the benefits of opportunities that flow as a result.

***

According to the executive director of the United Nations Environment Programme, Inger Andersen, nature is sending us a message via the corona virus pandemic and the ongoing climate crisis.[3] She said humanity was placing too many pressures on the natural world with damaging consequences, and warned that failing to take care of the planet means not taking care of ourselves. She asserted that nature is sending us a warning that if we neglect the planet, we put our own wellbeing at risk.

The corona virus outbreak was a "clear warning shot", given that far more deadly diseases exist in wildlife, and that today's civilisation is "playing with fire" as it is almost always human behaviour that causes diseases to spill over into humans.

Inger Andersen, Executive Director, United Nations Environment Programme. Nature is sending us a message via the corona virus pandemic.

Experts confirm that both global warming and the destruction of the natural world are responsible for driving wildlife into closer contact with people. They also urged authorities to put an end to the illegal global animal trade, and to live animal markets—which they called an "ideal mixing bowl" for disease. Inger Andersen asserts that never before have so many opportunities existed for pathogens to pass from wild and domestic animals to people, and that 75 per cent of all emerging infectious diseases come from wildlife.

There are too many pressures on our natural systems and something has to give. We are intimately interconnected with nature, whether we like it or not. If we don't take care of nature, we can't take care of ourselves. And as we hurtle towards a population of 10 billion people on this planet, we need to go into this future armed with nature as our strongest ally.[4]

In *The Sydney Morning Herald* on 7 May 2020[5] the editorial was clear that a corona virus recovery strategy must apply to climate change, saying that the pandemic had drawn attention away from climate change, but the threat global warming poses to our planet has certainly not gone away.

Rising global temperatures might not be making the front page as often as during Australia's summer bushfire crisis, but the evidence of the damage already being inflicted is growing more compelling by the day.

With the pandemic in Australia at least for now under control these dangers must come back into focus and our response to the Corona virus provides important lessons for how we should respond. Above all Australia should take the same evidence-based scientifically lead approach to climate change as we took to COVID-19.

It is time the federal government accepted the warnings from scientists about climate change just as seriously and took a similarly proactive approach. For too long the government has downplayed the science, sidelined certain experts and allow policy to be influenced by lobby groups, including the fossil fuel industry.

Certainly, it will be important to promote economic recovery, but the pandemic has shown how early action is often the most economically sensible approach and avoids damage later.

The pandemic has shown us that when faced with huge problems politicians who rely on expert advice and act on a non-partisan fashion can find solutions.

Australia must develop rational evidence-based climate change policies. We shouldn't jettison the lessons of our success against COVID-19 even before the crisis is over.

Human infectious disease outbreaks are increasing. In recent years there have been Ebola, bird flu, Middle East respiratory syndrome (MERS), Rift Valley fever, severe acute respiratory syndrome (SARS), West Nile virus and Zika virus. All have jumped the species barrier from animals to humans.

Renowned conservationist and activist Dr Jane Goodall is hoping the corona virus pandemic will be a wake-up call, warning the crisis is a result of human disregard for nature and animals. She considers this disregard for nature as equally the root cause of climate change.

Dr Goodall says the loss of animal habitats and intensive farming make it easier for viruses to spread from one animal to another and then to humans.[6] "Most of these viruses that jumped to us have come through an intermediary. So, there's a reservoir host like a bat, and in the case of COVID-19 it's thought to have jumped into a pangolin and then into us."

The Zoological Society of London's Professor Andrew Cunningham says the emergence and spread of the corona virus was not only predictable, it was predicted in the sense that another viral emergence from wildlife was expected to be a public health threat. He pointed to a study of the 2002/03 SARS outbreak that concluded: "The presence of

ABOVE LEFT: Dr Jane Goodall, renowned conservationist and activist. She hopes the corona virus pandemic will be a wake-up call, warning the crisis is a result of human disregard for nature and animals. ABOVE RIGHT: Prof Andrew Cunningham. The emergence and spread of the corona virus was not only predictable, it was predicted in the sense that there would be another viral emergence from wildlife that would be a public health threat.

Market butchering live animals, Indonesia. People in markets and in intimate contact with the animal body fluids make an ideal mixing bowl for disease emergence and transmission.

a large reservoir of Sars-CoV-like viruses in horseshoe bats, together with the culture of eating exotic mammals in southern China, is a timebomb."[7]

Cunningham believes that although we might not think so, "at the moment, we've probably got a bit lucky. . . So, I think we should be taking this as a clear warning shot. It's a throw of the dice."

"It's almost always a human behaviour that causes it and there will be more in the future unless we change. Markets butchering live wild animals from far and wide in China and elsewhere are the most obvious example," he added.

The animals have been transported over large distances and crammed together into cages. They are stressed and immunosuppressed and excreting whatever pathogens they have in them. With people in large numbers in the market and in intimate contact with the body fluids of these animals, you have an ideal mixing bowl for [disease] emergence. If you wanted a scenario to maximise the chances of [transmission], I couldn't think of a much better way of doing it.[8]

Not unsurprisingly, the environmental changes arising from the corona virus were first visible from space. As the virus-induced community and economic lockdown spread, changes were evident in the atmosphere as well as on the ground.[9] While the human toll from the pandemic mounted horrendously, nature it seemed, was increasingly able to breathe more easily.

After decades of relentlessly increasing pressure, the human impact on the planet suddenly lightened. By mid-March 2020 air traffic had halved, and in some countries road traffic fell by more than 70 per cent. As motorways cleared and factories closed, pollution levels shrank within days of lockdown.[10] First China, then Italy, the United Kingdom, Germany and dozens of other countries experienced temporary falls in carbon dioxide emissions of as much as 40 per cent, greatly improving air quality and reducing the risks of asthma, heart attacks and lung disease.

Hopes that humanity could emerge from the catastrophic pandemic into a healthier, cleaner world depend not on the short-term impact of the virus, but on the long-term political decisions made about what follows the trauma.

Key environmental indices, which have steadily deteriorated for more than half a century or more, have paused or improved. In China, the world's biggest source of carbon, emissions were down. Europe has also seen a significant reduction. Substantial falls in emissions are anticipated

Air pollution from a brown coal power station chimneys. Could the corona virus mark the start of a prolonged downward trend in emissions and the beginning of the end for fossil fuels?

in the United States of America where vehicle traffic—its major source of carbon emissions—has fallen by about 40 per cent. Even assuming a bounce back once the lockdowns lifteds, the planet is expected to see its first fall in global emissions since the 2008-2009 financial crisis. This is potentially good news for the climate. Some analysts believe the corona virus could mark the start of a prolonged downward trend in emissions and the beginning of the end for fossil fuels.

Chair of the Global Carbon Project Rob Jackson confirmed that the drop in emissions is global and unprecedented in scale. "Air pollution has plunged in most areas. The virus provides a glimpse of just how quickly we could clean our air with renewables."[11]

However, he added that the human cost had been high and the environmental gains could prove temporary. "I refuse to celebrate a drop in emissions driven by tens of millions of people losing their jobs. We need systemic change in our energy infrastructure, or emissions will roar back."

Interestingly, but not surprisingly, Chapter 17 described the resurgence of natural life in the "Zone of Alienation" and around the Chernobyl Nuclear Power Plant and the Demilitarized Zone in Korea. In a similar way, while humans were in temporary retreat during the corona virus lockdown, the whole planet became not only cleaner, but wilder. Wildlife began to fill the vacuum created. Fewer vehicles meant a staggering reduction in road kill that in the United Kingdom alone usually takes an annual toll of about 100,000 hedgehogs, 30,000 deer, 50,000 badgers and 100,000 foxes, as well as barn owls and other species.[12]

Normally timid, coyotes have been spotted on the Golden Gate

Bridge in San Francisco. Deer are grazing near Washington homes a few miles from the White House. Wild boar are becoming bolder in Barcelona and Bergamo, Italy. In Wales, peacocks have strutted through Bangor, and in Llandudno goats walked down streets that were deserted due to a virus-induced lockdown. Kangaroos have bounded unhindered on roads deserted by traffic in Australian cities.

Ultimately, the most important environmental impact of the corona virus pandemic is likely to be on public perceptions. The virus demonstrated the deadly consequences of ignoring expert warnings, of political delays, and of sacrificing human health and natural values for economic goals. Whether the pandemic is good or bad for the environment will depend not on the virus, but on humanity. If there is no political pressure on governments, the world will go back to unsustainable business as usual, rather than emerging with a healthier sense of what should be normal.

TOP: Goats walk the quiet streets in Llandudno, North Wales, deserted due to a corona virus-induced lockdown. ABOVE: French philosopher Bruno Latour. Humanity has learnt that it is possible in a matter of just weeks to slow the economy, which until now had been considered inconceivable due to the pressures of globalisation.

French philosopher Bruno Latour reckons humanity has learnt now that it is possible in a matter of just weeks to slow the economy, which until now had been considered inconceivable due to the pressures of globalisation.[13] "The incredible discovery is that there was in fact in the world economic system, hidden from all eyes, a bright red alarm signal, next to a large steel lever that each head of state could pull at once to stop 'the progress train' with a shrill screech of the brakes," he wrote.

This makes ecological calls to move off a path of endless resource consumption more realistic, and to heed them even more desirable.

SLEEPWALKING PAST A POINT OF NO RETURN

The generation that fiddled while the planet burned

"The world must choose hope over surrender in the fight against climate change," said United Nations Secretary-General António Guterres in December 2019, warning a summit in Madrid[1] that governments risked sleepwalking past a point of no return.[2]

The annual negotiations to bolster the 2015 Paris Agreement aimed at curbing global warming, held in Madrid in 2019, began against a backdrop of severe weather disasters, ranging from fires in the Arctic, the Amazon and Australia to intense tropical hurricanes.

"Do we really want to be remembered as the generation that buried its head in the sand, that fiddled while the planet burned?" Guterres asked the opening session.

"One is the path of surrender. The other option is the path of hope. A path of resolve, of sustainable solutions," he said.

The summit was also seen as a broader test of international commitments to major changes in energy, transport and industry that scientists say are needed to wean the world off fossil fuels quickly enough to avoid the

LEFT: United Nations Secretary-General António Guterres: "Do we really want to be remembered as the generation that buried its head in the sand, that fiddled while the planet burned?" Source: Middle East Online RIGHT: Pedro Sanchez, Spain's acting Prime Minister: "There is no wall high enough to protect any country, however powerful."

catastrophic impacts of climate change. Under existing greenhouse gas reduction pledges, the world is on course for the kind of temperature rises that could threaten the stability of industrial countries, and towards the end of this century lay waste to swathes of the developing world.

Spain's acting Prime Minister Pedro Sanchez, whose government hosted the summit of 27,000 delegates, urged attendees to take ambitious action to preserve the "fragile balance" of life on Earth. No one can independently pull out of this challenge, he told delegates.[3] "There is no wall high enough to protect any country from this challenge, however powerful they are."

Secretary-General Guterres told the Madrid summit that the world stood at a critical juncture in its collective efforts to limit dangerous global warming. He said the path of hope for the world was for more fossil fuels to remain where they should be—in the ground.

"That is the only way to limit global temperature rise to the necessary 1.5 degrees by the end of this century."

"Millions throughout the world—especially young people—are calling on leaders from all sectors to do more, much more, to address the climate emergency we face. They know we need to get on the right path today, not tomorrow. That means important decisions must be made now."

He pointed out that levels of greenhouse gases in the atmosphere had reached another record high. Global average levels of carbon dioxide reached 407.8 parts per million in 2018, beyond the 400 parts per million that was previously seen as an unthinkable tipping point. The last time there was a comparable concentration of carbon dioxide in the atmosphere was between 3 and 5 million years ago, when the temperature was between 2 and 3 degrees Celsius warmer than it is today, and sea levels were 10 to 20 metres higher than at present.

"We need a rapid and deep change in the way we do business, how we generate power, how we build cities, how we move, and how we feed the world. If we don't urgently change our way of life, we jeopardise life itself," said Guterres. "The only solution is rapid, ambitious, transformative action by all—governments, regions, cities, businesses and civil society, all working together towards a common goal. To put a price on carbon is vital if we are to have any chance of limiting global temperature rise and avoiding runaway climate change."

"We must finally demonstrate that we are serious in our commitment to stop the war against nature—that we have the political will to reach carbon neutrality by 2050. There is no time and no reason to delay. We have the tools, we have the science, we have the resources. Let us show we also have the political will that people demand from us. To do anything less will be a betrayal of our entire human family and all the generations to come."

At the conclusion of the summit Secretary General Guterres said he was "disappointed" with the results and that "the international community lost an important opportunity to show increased ambition on mitigation, adaptation and finance to tackle the climate crisis."

Although the world's major emitters weren't expected to announce fresh climate pledges at the Madrid summit, there was still hope that they might collectively send a strong message of intent for next year. However, talks quickly became bogged down in technical issues.

There was a growing sense among many attending the Madrid summit of a disparity between these slow, impenetrable United Nations processes and the action being demanded by protesters around the world. This was summarised by the executive director of Greenpeace Jennifer Morgan,[4] who said that despite the "fresh momentum" provided by the growing global climate movement, it was yet to penetrate the halls of power.

Capturing global headlines in 2019, Swedish teenage environmental activist Greta Thunberg spoke at the United Nations on 23 September 2019 about climate change, accusing world leaders of inaction and half measures.[5]

Swedish teenage environmental activist Greta Thunberg, who accuses world leaders of inaction and half measures.

You have stolen my dreams and my childhood with your empty words and yet I'm one of the lucky ones. People are suffering. People are dying. Entire ecosystems are collapsing. We are in the beginning of a mass extinction and all you can talk about is money and fairy tales of eternal economic growth. How dare you!

For more than 30 years, the science has been crystal clear. How dare you continue to look away and come here saying that you're doing enough when the politics and solutions needed are still nowhere in sight.

You are failing us, but the young people are starting to understand your betrayal. The eyes of all future generations are upon you and if you choose to fail us, I say: We will never forgive you.

We will not let you get away with this. Right here, right now is where we draw the line. The world is waking up and change is coming, whether you like it or not.

A common refrain from protesters and observers at the Madrid summit was the discrepancy between the slow pace of the talks and the urgency suggested by the latest science. Arriving at the summit Greta Thunberg summarised the mood when she told those assembled in the main hall that the summit "seems to have turned into some kind of opportunity for countries to negotiate loopholes".

Shortly after her appearance, she was announced as *Time Magazine's* Person of the Year.

Prior to the Madrid summit the report *World Scientists' Warning of a Climate Emergency* published in the journal *Bioscience* in November 2019 by a team of more than 11,000 scientists warned that the planet "clearly and unequivocally faces a climate emergency." The report's signatories included scientists from 153 countries, collectively known as the Alliance of World Scientists. The report warns that urgent action is needed "to avoid untold suffering due to the climate crisis" if deep and lasting shifts in human activities that contribute to climate change are not made. It warns that climate change is intensifying faster than most scientists predicted and is "threatening natural ecosystems and the fate of humanity".

The report is based on 40 years of data that show how human activities have affected the environment, including changes in fossil fuel consumption, carbon dioxide emissions, rates of deforestation and global surface temperatures. It says that scientists have "a moral obligation to clearly warn humanity of any catastrophic threat."

Scientists have collectively used the world "emergency" in reference to climate change. The report says "climate change has arrived, and is accelerating faster than many scientists expected."

The report argues that family planning services and other social justice efforts that promote full gender equity should be enacted to help stabilise the world's population, which is increasing by approximately 80 million per year. Also, that countries should replace fossil fuels with renewable energy sources, while investing in technologies to extract carbon dioxide from the atmosphere. Further, that governments should end subsidies to fossil fuel companies, and that wealthier countries should support poorer nations in transitioning to cleaner energy sources.

It adds that nations must sharply reduce emissions of atmospheric pollutants, including compounds that are commonly used in air conditioning, refrigeration and aerosols. Researchers say that reducing these short-lived pollutants could slow the planet's short-term warming trend by more than 50 per cent over the next few decades.

The report advocates climate change mitigation efforts that focus on protecting and restoring ecosystems such as forests, coral reefs, savannas and wetlands, which naturally absorb and store carbon dioxide. Also, that

Restored tropical forest. Climate change mitigation efforts should focus on protecting and restoring ecosystems such as forests, coral reefs, savannas and wetlands, which naturally absorb and store carbon dioxide. Source: John Halkett

Supporters of the Extinction Rebellion climate change movement in London. A grass roots, citizen-based movement, it uses civil disobedience in an attempt to "halt mass extinction and minimise the risk of social collapse".

people should be encouraged to eat mostly plant-based food, which will improve health and lower greenhouse gas emissions from livestock. Also, that economies should prioritise carbon-free initiatives and sustaining eco-systems, rather than focusing on GDP growth and the pursuit of increasing affluence.

The scientists said it will take strong actions by the public to move politicians toward adopting lasting policy changes. The report applauds "grassroots citizen movements" and youth-led movements calling for climate action. "We believe that the prospects will be greatest if decision-makers and all of humanity promptly respond to this warning and declaration of a climate emergency, and act to sustain life on planet Earth, our only home."

University of Sydney ecologist Dr Thomas Newsome says these measures should become part of the public discussion of climate change.[6] "There have been plenty of people willing to criticise these kids that perhaps they don't understand the science, but it's increasingly clear that a lot of the youth leading these protests understand the science much better than any of us. It was important for us scientists to say—yes, the situation is that dire."

His observations include the relatively radical movement Extinction Rebellion, a grassroots, citizen-based movement, which uses civil disobedience in an attempt to "halt mass extinction and minimise the risk of social collapse".

"Everyone has a right to peaceful protest. So, I guess people are just looking for another way to get this message across, partly because governments haven't been acting and responding in a way that is their duty."

TOWARDS THE CLIFF EDGE

Determined to be the masters of our own demise

All the pointers indicate that humanity is close to the edge of the cliff—marching onwards towards oblivion. We know it, but nevertheless march on we do. Having established ourselves as masters of the Earth, we now seem determined to be the masters of our own demise.

🌳🌳🌳

Just because my views about Christian churches are not positive, doesn't mean I can't draw on the Bible for a bit of historical and moral direction. According to the Bible, Jesus sent his angel to make what was coming known to his servant John and John wrote it down in the book of Revelation, swearing that it was the word of God guaranteed by Jesus Christ. "Happy the man," wrote John, "who reads this prophecy and happy those who listen to him if they treasure all that it says because the Time is close."

John goes on to say: "The time has come for your own anger and for the dead to be judged and for your servants the prophets, for the saints and for all who worship you, small or great, to be rewarded. The time has come to destroy those who are destroying the Earth."[1]

Ice-free northern Arctic sea. The worst possible case for an ice-free Arctic summer, predicted for 2050, has now been brought forward by a couple of decades.

Many readers take that with a grain of salt, as the conventional interpretation deserves. But in the context of the climate change debate these words are profound. Isn't John saying "So, we have a moral obligation to care for [the Earth], and to live as if our very lives and futures all depended on it."? We ignore that message our peril. We don't have a Planet B.

Scientists rightly caution that obviously some conclusions remain speculative. However, we can be sure that mounting violent weather events add up to a trend that means the climate has reached or passed a tipping point—that we here on Earth are in deep trouble. Scientists assert that if we wait to act until all the numbers are in, we will have waited too long. This is why scientists keep jamming every possible variable into models that predict our likely alarming climate future.

Because some conclusions are tentative, their credibility is attacked by whoever profits from business-as-usual, including some media shock jocks. But as earlier chapters affirm, the main failing to date in climate change models has been their timidity. For instance, the worst possible case for an ice-free Arctic summer, predicted for 2050 back in 2008, has now been brought forward by at least a couple of decades.

At what point might politicians, decision makers, influencers and carry-on-regardless industry captains be convinced that drastic climate change is already upon us, and likely to worsen—perhaps fatally—if we don't respond now and decisively. Too late already? So just carry on over the edge of the cliff is firming up as an odds-on bet.

When it comes to picking winners and losers in terms of humanity's willingness to sacrifice fellow animal travellers on planet Earth, selecting exactly which ones we can or cannot live without turns out to be an impossible gamble, in the grand experiment called life on Earth, because there is no control group. We will not know for certain until some species are gone for good, when it will be too late to call them back—sorry we were silly—we didn't really mean it—please come back!

Robust science-based on the best evidence available[2] identifies boundaries, beyond which the world will enter a phase shift that is likely to prove cataclysmic for humanity. Predictable breaches of the boundaries identified by the scientific analysis include those that are critical to human survival, amongst them climate change, biodiversity loss, disruption of global nitrogen and phosphorus cycles, ozone depletion, ocean acidification, and fresh water scarcity. All of the indicators on these are now pointing in the wrong direction.

Biodiversity loss is obviously a major issue. Before the Industrial Revolution, the fossil record suggests 0.1 to 1 species per million became extinct annually. The acceptable limit is suggested to be 10. It's a bit like the road toll—how many fatalities are acceptable? The actual current loss is at least 100 species each year out of every million, a figure widely feared to rise tenfold this century. Nothing remotely similar has happened since that asteroid in Chapter 3 did it in for the dinosaurs.

African girls struggling to pump water out of an almost dry public well. Fresh water scarcity threatens human survival.

🌳 🌳 🌳

A landmark report in 2019[3] warned that as many as one million species of animals, insects and plants are threatened with extinction in the coming decades. A more recent study found that the growth of cities, the clearing of forests for farming and the soaring demand for fish has significantly altered nearly three-quarters of the land and more than two-thirds of the oceans.

One of the scientists involved, Professor Andy Purvis of the Natural History Museum in London, said that by undermining important habitats; "we're hacking away at our safety net, we're trashing environments we depend on."[4]

He pointed to the impact of everything from the use of palm oil in processed food and shampoo to the pressures created by fast fashion. While the need for conservation is understood in many developed countries, Professor Purvis said: "We've exported the damage to countries too poor to handle the environmental cost of what they're selling to us."

I mentioned David Attenborough in chapter 14 and referred to his 2020 book, *A Life on Our Planet: My Witness Statement and a Vision for the Future.* In relation to his deep concerns about climate change and the lack of meaningful action, he said that humanity's impact on the climate is now truly global.

"Our blind assault on the planet is changing the very fundamentals of the living world. This is now the status of our planet in the year 2020 … We have replaced the wild with the tame. We regard the earth as our planet run by humankind for humankind. There is little left for the rest of the living world. The truly wild world—that nonhuman world—has gone. We have overrun the Earth … that will bring about nothing less than the collapse of the living world the very thing that our civilisation relies on."[5]

Sir David details that the latest in scientific understanding that suggests the living world is on course to tip and collapse. Indeed, he argues it has already begun to do so, and is expected to continue with increasing speed, such that the effects of its decline will become greater in scale and more impactful as they follow one after the other.

"Everything we have done come to rely upon—all the services that the Earth's environment has always provided us for free could begin to falter or fail entirely."[6]

"We share it [the Earth] with the living world—the most remarkable life-support system imaginable, constructed over billions of years to refresh and renew food supplies, to absorb and reuse waste, to dampen damage and bring balance at the planetary scale. It is no accident that the planet's stability has wavered just as its biodiversity has declined—the two things are bound together. To restore stability to our planet, therefore, we must restore its biodiversity. It is the only way out of this crisis that we ourselves have created."[7]

In a final conciliatory note Sir David said there may still be time to make amends, to manage our impact, change the direction of our development and once again become a species in harmony with nature; "All we require is the will. The next few decades represent a final opportunity to build a stable home for ourselves and restore the rich, healthy, wonderful world that we inherited from distant ancestors. Our future on the planet, the only place as far as we know where life of any kind exists, is at stake."[8]

It will now take a Herculean effort to pull humanity back from the edge of that precipice. I do not detect a willingness or the political courage to make this effort. Essentially business-as-usual, with some fiddling at the edges, will all but guarantee we will vacate the planet, allowing trees and nature to reassert themselves—and perhaps over geological time a smarter breed of bipeds may inherit the Earth.

Reflecting on Attenborough's words, where to from here for humanity's future? Chapter 17 outlined some of the benefits of a return of nature for the planet, if we humans were to disappear. Bruised and battered nature would come rushing back, but the narrative left open the prospect that after the planet's next great extinction there might still be a life for future human generations. What would that look like?

Really, as Sir David concludes, a massive effort is needed urgently to pull humanity back from the edge of the cliff. But there is hope. Unlike other animals including dinosaurs, mammoths and every other species that evolved over multiple generations, humanity has demonstrated the capacity to adapt culturally. This has enabled human communities to adjust lifestyles and to develop technological solutions to rapidly cope rapidly with altered environmental conditions. No other group of living organisms has demonstrated such a capacity. Limited to gradual biological adaptations, dinosaurs and mammoths could not alter their ways of life quickly enough to avoid extinction.

So, in theory at least, after any major environmental or self-induced catastrophe, as long as a nucleus of breeding humans remains with access to air, water, shelter and food, the human race is unlikely to become extinct, unless those survivors fail to compete with other species.

What might it be like for this remerging post-apocalyptic human population? The resurgence of nature during the corona virus pandemic, as outlined in the previous chapter, provides a hint. A post-apocalyptic

Chimpanzee group meeting. Human successors in the form of an intelligent biped may emerge here on Earth—hopefully they prove to be a bit smarter than us.

human society would not have the burden of serious and escalating over-population that challenges humanity today. So, more nature and fewer people would be part of a recipe for a sustainable human race across the generations to come.

Rather than abrogate the planet entirely to a resurgent nature, a future human society may well have learnt the lessons of its predecessors, and be much more receptive to sustainable lifestyles that acknowledge the relationship between humanity and nature. Material assets and the holy grail of financial affluence may be relegated down the order in favour of a frugal existence and a more harmonious relationship with the natural world.

Future generations will need to turn their backs on the indulgences of present generations, eating less meat and pursuing more modest expectations. They will have to make do with sustainable resources and energy, and recognise the critical equilibrium between future survival and natural systems. So there will be no plastics, no synthetics and no other products requiring high energy manufacturing. A back-to-basics approach, including reliance on renewable products like wood, and placing trees in a predominant position in relation to the prospect of a durable, long-term survival for the human race.

Certainly, the planet of the future will be a planet of the trees. Trees are mobile, adaptable and opportunists ready to exploit favourable opportunities—when humans draw back trees will move in! Without humans, or with fewer of them, and with more trees over a significant period of time atmospheric conditions will moderate—temperatures will come back down, atmospheric carbon levels will decline and conditions for emergent, better adjusted human generations or another form of intelligent bipeds will emerge here on Earth—hopefully a bit smarter than we have proved to be.

If humanity is to stare down its prospects of following the dinosaurs, Neanderthals and other life forms into extinction, urgent action is needed. A willingness to undertake such action is yet to be universally acknowledged and implemented. Strong international consensus and leadership will be needed if survival prospects for humanity are to be improved. However, in my view, there is little evidence that this will happen in the near future.

NOTES

PREFACE

1. United Nations General Assembly, 26 September 2019.
2. *Climate of the Nation*, The Australia Institute, 2018 and *Australian Survey into Beliefs and Attitudes towards Science*, National Centre for Public Awareness of Science, 2018.

CHAPTER 1: PREAMBLE

1. *New York Times*, 3 January 2020. Novelist Richard won the 2014 Man Booker Prize for *The Narrow Road to the Deep North*. Each of his works has been critically acclaimed and won awards.
2. Grahame Madge, "Australian Bushfires Help Push Forecast 2020 CO$_2$", British Met Office, 24 January 2020.
3. "Earth's carbon dioxide levels hit record high, despite coronavirus-related emissions drop", *The Washington Post*, Andrew Freedman and Chris Mooney, 4 June 2020.
4. Temporary reduction in daily global CO$_2$ emissions during the COVID-19 forced confinement, 2020, e Quéré, C., Jackson, R.B., Jones, M.W., et al, *Nat. Clim. Chang.* 10, 647–653. https://doi.org/10.1038/s41558-020-0797-x

CHAPTER 2: SETTING THE SCENE

1. "The Greenland ice sheet poured 197 billion tons of water into the North Atlantic in July alone", Andrew Freedman and Jason Samenow, *The Washington Post*, 4 August 2019. See: https://www.washingtonpost.com/weather/2019/08/01/greenland-ice-sheet-poured-billion-tons-water-into-north-atlantic-july-alone/

CHAPTER 3: ENCOUNTERS FROM OUT THERE IN SPACE

CHAPTER 4: IT'S HEATING UP HERE ON PLANET EARTH

1. "Pall of Smoke a Grim Reminder of Climate Risks", Editorial, *The Sydney Morning Herald*, 23–24 November 2019, p 32.
2. "How Cold is Pluto?" Nola Taylor Redd. *Space.com* article, 20 February 2016. See: https://www.space.com/18563-pluto-temperature.html
3. "What Is the Sun's Role in Climate Change?" *NASA's Global Climate Change Website* Blog, 6 September 2019. See: https://climate.nasa.gov/blog/2910/what-is-the-suns-role-in-climate-change/
4. "Carbon dioxide at highest level in 650,000 years", *Mongabay.com*, 24 November 2005. See: https://news.mongabay.com/2005/11/carbon-dioxide-at-highest-level-in-650000-years/
5. "Predictions of Future Global Climate", *UCAR Center for Science Education*, 2020. See: https://scied.ucar.edu/learning-zone/impacts-climate-change/predictions-future-global-climate
6. "Glacial-Interglacial Cycles", *National Centers for Environmental Information*, National Oceanic and Atmospheric Administration. See: https://www.ncdc.noaa.gov/abrupt-climate-change/Glacial-Interglacial%20Cycles
7. "Sea level rise, explained", Christina Nunez, *National Geographic*, 20 February 2019. See: https://www.nationalgeographic.com/environment/article/sea-level-rise-1
8. "If we stopped emitting greenhouse gases right now, would we stop climate change?", *The Conversation*, 8 July 2017. See: https://theconversation.com/if-we-stopped-emitting-greenhouse-gases-right-now-would-we-stop-climate-change-78882
9. "Forests and Climate Change",

Issues Brief, International Union for Conservation of Nature, February 2021. https://www.iucn.org/sites/dev/files/forests_and_climate_change_issues_brief.pdf

10. This organisation was established in 1988 by the United Nations Environment Programme and the World Meteorological Organization to provide a scientific view on the current state of knowledge in climate change and its potential environmental and socio-economic impacts.

11. "Recent Pause in the Growth Rate of Atmospheric CO_2 Due to Enhanced Terrestrial Carbon Uptake", Trevor F Keenan, Colin Prentice, Josep G Canadell, Christopher A Williams, Han Wang, Michael Raupach and James Collatz. *Nature Communications*, 8 November 2016.

12. "If You're Looking For Good News About Climate Change, This is About the Best There is Right Now", Chris Mooney. *The Washington Post,* 10 November 2016.

13. "Arctic Sea Ice Minimum", *NASA Global Climate Change: Vital Signs of the Planet.* 16 December 2020. See: https://climate.nasa.gov/vital-signs/arctic-sea-ice/

14. "Sea Level Rise and Implications for Low-Lying Islands, Coasts and Communities", *Special Report: Special Report on the Ocean and Cryosphere in a Changing Climate*, Chapter 4, International Panel on Climate Change. 2016. See: https://www.ipcc.ch/srocc/chapter/chapter-4-sea-level-rise-and-implications-for-low-lying-islands-coasts-and-communities/

CHAPTER 5: DROUGHT, FLOODS AND BUSHFIRES

1. "Australia's Dry Run", Robert Draper, *National Geographic*, April 2009.

2. "Darwin's Experiences in Australia", Paper presented by Emeritus Professor Frank Nicholas, University of Sydney, *Darwin Symposium*, National Museum of Australia, 26 February 2009. See: https://www.nma.gov.au/audio/charles-darwin-series/transcripts/darwins-experiences-in-australia

3. *Australia in its Physiographic and Economic*

Aspects, (Clarendon Press: 1911), Thomas Griffith Taylor, Kessinger Publishing: February 2010.

4. The Palmer Drought Severity Index (PDSI) uses readily available temperatures and rainfall data to estimate relative dryness. It is effective in determining long-term drought, especially over low and middle latitudes.

5. Droughts and Floods in Australia: The Big Dry and other iconic droughts. See: https://ummenhofer.whoi.edu/southeast-australian-drought/ Caroline Ummenhofer Lab Studying climate and rainfall variability around the globe. Woods Hole Oceanographic Institute, 2014.

6. Indian Ocean Variability & Regional Impacts. See: https://ummenhofer.whoi.edu/indian-ocean-variability-regional-impacts/ Caroline Ummenhofer Lab Studying climate and rainfall variability around the globe. Woods Hole Oceanographic Institute, 2014.

7. A global dataset of "Palmer Drought Severity Index for 1970-2002 Relationship with Soil Moisture and Effects of Surface Warming", Dai A Trenberth, KE, and Qian, T., *Journal of Hydrometeorology* 5, 2004, p 1117–30.

CHAPTER 6: LEARNING TO LIVE IN A GREENHOUSE

1. "CO2 emissions", Hannah Ritchie and Max Roser, *Our World in Data*, 2018. See: https://ourworldindata.org/co2-emissions

2. "Africa is particularly vulnerable to the expected impacts of global warming". United Nations, *Fact Sheet on Climate Change*. 2006. See: https://unfccc.int/files/press/backgrounders/application/pdf/factsheet_africa.pdf

3. "Climate Change 2007: Synthesis Report", Intergovernmental Panel on Climate Change, 2007.

4. For further information see the Great Barrier Reef Marine Park Authority's article on coral bleaching at: https://www.gbrmpa.gov.au/the-reef/reef-health/coral-bleaching-101

5. "1.06 C Above 1880: Climate Year 2015 Shatters All Previous Records For Hottest Ever", *Robertscribbler.*

com, 14 December 2015. See: https://robertscribbler.com/2015/12/14/1-06-c-above-1880-climate-year-2015-shatters-all-previous-records-for-hottest-ever-recorded/?cpage=1

CHAPTER 7: DID WE KILL OFF OUR COUSINS?

1. "The Things That Divide Us", David Berreby, *National Geographic: The Race Issue*, April 2019, p 52.
2. "Rwanda genocide: 100 days of slaughter", *BBC News*, April 2019. https://www.bbc.com/news/world-africa-26875506
3. "Humans' Ancestral 'Homeland' Traced to Northern Botswana", Martin Macias Jr., *Courthouse News Service*, 28 October 2018. See: https://www.courthousenews.com/humans-ancestral-homeland-traced-to-northern-botswana/
4. "Palynomorphs from a sediment core reveal a sudden remarkably warm Antarctica during the middle Miocene", Sophie Warny, Rosemary Askin, et al., *Geology* 37(10), The Geological Society of America, October 2009, 955–958.
5. "The Great Human Migration: Why humans left their African homeland 80,000 years ago to colonize the world", Guy Gugliotta, *Smithsonian Magazine*, July 2008.
6. "The Great Human Migration: Why Humans Left Their African Homeland 80,000 Years Ago to Colonize the World", Guy Gugliotta, *Smithsonian Magazine*, July 2008. See: https://www.smithsonianmag.com/history/the-great-human-migration-13561/
7. "40,000 Year Old Evidence that Neanderthals Wove String", CNRS, *Science Daily*, 9 April 2020. https://www.sciencedaily.com/releases/2020/04/200409110533
8. "Neandertals, Stone Age People May Have Voyaged the Mediterranean", Andrew Lawler, *Science Magazine*, 24 April 2018. See: https://www.sciencemag.org/news/2018/04/neandertals-stone-age-people-may-have-voyaged-mediterranean
9. "Humans and Neanderthals had sex, but not very often", Ed Young. *National Geographic*, 12 September 2011. See: https://www.nationalgeographic.com/science/article/humans-and-neanderthals-had-sex-but-not-very-often
10. "Neanderthals, Humans Interbred: First Solid DNA Evidence", Ker Than. *National Geographic News*, 8 May 2010. See: https://www.nationalgeographic.com/news/2010/5/100506-science-neanderthals-humans-mated-interbred-dna-gene/

CHAPTER 8: PLANET EARTH—WE HAVE A PROBLEM

1. "World Population Growth", Max Roser, Hannah Ritchie and Esteban Ortiz-Ospina in *Our World in Data,* May 2019. https://ourworldindata.org/world-population-growth
2. "The Growing World Population", *Population Summit of the World's Academies*, The National Academies Press, Washington DC, USA, 1993.
3. "Growing at a slower pace, world population is expected to reach 9.7 billion in 2050 and could peak at nearly 11 billion around 2100", United Nations, Department of Economic and Social Affairs, 17 June 2019. See: https://www.un.org/development/desa/en/news/population/world-population-prospects-2019.html
4. "Growing at a slower pace, world population is expected to reach 9.7 billion in 2050 and could peak at nearly 11 billion around 2100", United Nations, Department of Economic and Social Affairs, 17 June 2019. See: https://www.un.org/development/desa/en/news/population/world-population-prospects-2019.html
5. *World Population Prospects Data Booklet.* United Nations, Department of Economic and Social Affairs, 17 June 2019. See: https://population.un.org/wpp/Publications/Files/WPP2019_DataBooklet.pdf
6. Ibid.
7. *The World Bank in Pakistan*, The World Bank, April 2020. www.worldbank.org/en/country/pakistan/overview
8. "World Population Growth", Max Roser, Hannah Ritchie and Esteban Ortiz-Ospina in *Our World in Data,* May 2019. https://ourworldindata.org/world-

population-growth

9. "Why Population Matters", The Issue, *Population Matters*, 2020. See: https://populationmatters.org/the-issue

10. "Resources & Consumption", The Facts, *Population Matters*, 2020. See: https://populationmatters.org/the-facts/resources-consumption

11. "In 1850, Ignaz Semmelweis saved lives with three words: wash your hands", Dr Howard Markle. *PBSO News Hour*, 15 May 2015. See: https://www.pbs.org/newshour/health/ignaz-semmelweis-doctor-prescribed-hand-washing

12. "Powder of the Devil' ... the Revolutionary Cure for Malaria", John Halkett, *Talking Trees*, 24 September 2019. See: http://www.talkingtrees.com.au/the-blog/

13. "This Army doctor made the Panama Canal possible by killing mosquitoes", Logan Nye, *The Mighty*, 2 April 2018. See: https://www.wearethemighty.com/articles/this-army-doctor-made-the-panama-canal-possible-by-killing-mosquitoes/

14. "Florey Howard Walter (Baron Florey) (1898–1968)" in the *Australian Dictionary of Biography*, vol 14, Melbourne University Press, pp 188–190.

15. *The Legacy: An Elder's Vision for our Sustainable Future*, David Suzuki, Allen & Unwin: Vancouver, 2013. p 15.

CHAPTER 9: NATURE'S LIGHT SHOW

1. "What Was It Like When Oxygen Appeared And Almost Murdered All Life On Earth?" Ethan Siegel, *Forbes*, March 2019. See: www.forbes.com/sites/startswithabang/2019/03/20/what-was-it-like-when-oxygen-appeared-and-almost-murdered-all-life-on-earth

2. "Venus: The hot, hellish & volcanic planet", Charles O Choi, *Space.com*, 16 January 2020. See: www.space.com/44-venus-second-planet-from-the-sun-brightest-planet-in-solar-system

CHAPTER 10: GETTING TO GRIPS WITH TREES

1. *The World of the Kauri*, John Halkett and EV Sale, Reed Methuen: Auckland, 1986.

2. Carbon 14 is radioactive carbon that continuously forms in the atmosphere and then gradually decays. This means that the ratio of carbon 14 to other carbon in the atmosphere is always the same. However, once carbon 14 is incorporated into inactive biomasses, for instance wood, the process of decay continues unabated, but no new radioactive carbon is accumulated. The lower the amount of radioactive carbon it contains, the older is the tissue.

3. "World's Oldest Living Tree – 9550 years old – Discovered In Sweden", *Science Daily*, 16 April 2008. See: https://www.sciencedaily.com/releases/2008/04/080416104320.htm

4. "Antarctica Forests", John Halkett. *Talking Trees*, 12 March 2018. See: http://www.talkingtrees.com.au/the-blog/

5. "Mission to gather petrified Antarctic plants could help predict future of flora on warming Earth", College of Liberal Arts & Sciences, University of Kanas. 2 June 2017. See: https://news.ku.edu/2017/12/01/mission-gather-petrified-antarctic-plants-could-help-predict-future-flora-warming-earth

CHAPTER 11: AT THE CENTRE OF HUMAN CONSCIOUSNESS

1. "Ancient Egypt", *Plant Explorers* online feature article. See: https://www.plantexplorers.com/explorers/history/index.html

2. "Uncrowned King of Trees", John Halkett, *Talking Trees*, 2 May 2017. See: http://www.talkingtrees.com.au/the-blog/

3. Charles Macintosh, one of the great Scottish inventors, was born into a family of prosperous Glasgow merchants on December 29, 1766. "On this day: Charles Macintosh, inventor of waterproofs, born", *The Scotsman*, 9 January 2017. See: https://www.scotsman.com/whats-on/arts-and-entertainment/day-charles-macintosh-inventor-waterproofs-born-1459340

4. "Charles Goodyear and the Vulcanization of Rubber", Ann Marie Somma, *Connecticut History.org*, 29 December 2014. See: https://connecticuthistory.org/charles-goodyear-and-the-vulcanization-of-rubber

5. "Rubber … from the Spanish Court to Pneumatic Tyres", John Halkett, *Talking Trees*, September 2019. See: www.talkingtrees.com.au/the-blog
6. Cork oak (*Quercus suber*)
7. "Everything You Could Ever Want To Know About Cork Trees", Vanya Maplestone, *Cycling Centuries*, 5 July 2019. See: https://www.cycling-centuries.com/everything-you-want-to-know-about-cork
8. *Around the World in 80 Trees*, Jonathan Drori, Laurence King Publishing: London, 2018, p 40.
9. "Hundreds flock to National Arboretum as Canberra's cork oak forest turns 100", Sherryn Groch, *The Canberra Times*, 13 November 2017.
10. Balsa (*Ochroma pyramidale*)

CHAPTER 12: FORESTS—WHERE TREES LIVE

1. "A Working Forest", Roger McDonald (ed)., *Gone Bush*, Bantam Books, 1990, pp 29–47.
2. Ibid.
3. "Trees grow almost anywhere—take the Joshua tree", John Halkett, *Talking Trees,* September 2019. See: www.talkingtrees.com.au/the-blog
4. "The State of the World's Forests 2020", *Forests, Biodiversity and People*, Food and Agriculture Organization of the United Nations, Rome. See: http://www.fao.org/documents/card/en/c/ca8642en
5. Broadleaved trees are flowering plants, that is they have 'normal' leaves as opposed to conifers.
6. *Gum*, Ashley Hay, Duffy & Snellgrove: Sydney, 2002.
7. "The State of the World's Forests 2020", *Forests, Biodiversity and People*, Food and Agriculture Organization of the United Nations, Rome. See: http://www.fao.org/documents/card/en/c/ca8642en
8. "Greenhouse gas emissions from tropical forest degradation: an underestimated source", Timothy R. H. Pearson, Sandra Brown, Lara Murray and Gabriel Sidman, *Carbon Balance and Management*, February 2017. See: https://cbmjournal.biomedcentral.com/articles/10.1186/s13021-017-0072-2
9. "CO_2 and Greenhouse Gas Emissions", Hannah Richie and Max Roser, *Our World in Data*, December 2019. See: https://ourworldindata.org/co2-and-other-greenhouse-gas-emissions
10. "Contribution of forest wood products to negative emissions: historical comparative analysis from 1960 to 2015 in Norway, Sweden and Finland", Cristina Maria Iordan, Xiangping Hu, Anders Arvesen, Pekka Kauppi and Francesco Cherubini, *Carbon Balance and Management*, 4 September 2018. See: https://cbmjournal.biomedcentral.com/articles/10.1186/s13021-018-0101-9

CHAPTER 13: A LANGUAGE THAT THE STRANGERS DON'T KNOW

1. "They speak a language that the strangers do not know", John Halkett, *Talking Trees*, September 2019. See: www.talkingtrees.com.au/the-blog
2. "Slow Wave Potentials — a Propagating Electrical Signal Unique to Higher Plants", Rainer Stahlberg, Robert Cleland, and Elizabeth Van Volkenburgh, *Communication in Plants: Neuronal Aspects of Plant Life*, University of Washington, January 2006. See: https://www.researchgate.net/publication/287916562_Slow_Wave_Potentials_-_a_Propagating_Electrical_Signal_Unique_to_Higher_Plants
3. "Net transfer of carbon between ectomycorrhizal tree species in the field", Suzanne W. Simard, David A. Perry, Melanie D. Jones, David D. Myrold, Daniel M. Durall and Randy Molina, *Nature*, Vol 388, 1997, pp 579–82.
4. "Death of a landscape: why have thousands of trees dropped dead in New South Wales?" Cris Brack and Catherine Ross, Australian National University, *The Conversation*, 15 October 2015. See: https://theconversation.com/death-of-a-landscape-why-have-thousands-of-trees-dropped-dead-in-new-south-wales-48657
5. Ibid.
6. "Snow gum dieback threatens to leave alpine waste land", Ricky French, *The Weekend Australian*, 5–6 October 2019 and "Dying Shame", Ricky French, *The Weekend Australian*, 5–6 October 2019. p 22–29.

7. Ibid.
8. Ibid.

CHAPTER 14: PUTTING A PRICE ON NATURE

1. *Green Capital: In A New Perspective on Growth*, Christian de Perthuis and Pierre-Andre Jouvet, Columbia University Press: New York, 2015, p 67.
2. "Conceptual model to support natural capital accounting of a forestry enterprise, a report from the Lifting Farm Gate Profits: The Role of Natural Capital Accounts project", AP O'Grady, EA Pinkard, RE Mount, RK Schmidt, ID Cresswell and SB Stewart, *CSIRO Land and Water*, May 2020.
3. "Existence of an Equilibrium for a Competitive Economy", Kenneth J. Arrow and Gerard Debreu via The Econometric Society, *Econometrica* Vol. 22, No. 3, 1954, pp 265–90.
4. "Definition of 'Pareto's Efficiency", *The Economic Times*, 4 January 2021. See: https://economictimes.indiatimes.com/definition/paretos-efficiency
5. "*The Tragedy of the Commons*", Garrett Hardin. *Science New Series*, Vol. 162, No. 3859, 13 December 1968, pp 1243–48.
6. An externality is a consequence of a commercial activity which affects other parties without this being reflected in market prices, such as the pollination of surrounding crops by bees kept for honey. So a cost borne by society, but not by a specific user is an externality.
7. *A Life on Our Planet: My Witness Statement and a Vision for the Future*, David Attenborough, Witness Books: Pelican Random House, 2020, p 178.

CHAPTER 15: SELLING CARBON IN THE MARKET

1. "How Much Do Health Impacts from Fossil Fuel Electricity Cost the US Economy?", Justin Gerdes, *Forbes*, 8 April 2013.
2. "The Climate Casino: Risk, Economics and Uncertainty for a Warming World", William Nordhaus, *Yale University Press*, 2013, p 177.
3. "These Countries Have Prices on Carbon. Are They Working?", Brad Plumber and Nadja Popovich, 2 April 2019, *The New York Times*. See https://www.nytimes.com/interactive/2019/04/02/climate/pricing-carbon-emissions.html
4. "British Columbia's Carbon Tax Yields Real World Lessons", Eduardo Porter, *New York Times*, 2 March 2016.
5. "The Economics of Tail Events with an Application to Climate Change", William Nordhaus. *Review of Environmental Economics and Policy*, Vol 5, Issue 2, 2011, pp 240–57.

CHAPTER 16: THAWING UP IN THE ARCTIC

1. "The Arctic Is Warming Twice as Fast as The Rest of The Planet", Bec Crew. *Science Alert*, December 2016. See: www.sciencealert.com/the-arctic-is-warming-twice-as-fast-as-the-rest-of-the-planet
2. "Arctic Sea Ice Minimum", in *Global Climate Change: Vital signs of the Planet*, via NASA, 2020. See: https://climate.nasa.gov/vital-signs/arctic-sea-ice/
3. "A thawing Arctic is heating up a new Cold War", Neil Shea, *National Geographic*, August 2018. See: https://www.nationalgeographic.com/adventure/2019/08/how-climate-change-is-setting-the-stage-for-the-new-arctic-cold-war-feature/
4. "The Arctic's thawing ground is releasing a shocking amount of dangerous gases", Craig Welch, *National Geographic*, 5 February 2020. See: https://www.nationalgeographic.com/science/2020/02/arctic-thawing-ground-releasing-shocking-amount-dangerous-gases/
5. "The irreversible emissions of a permafrost 'tipping point'", Christina Schädel, *The World Economic Forum*, 18 February 2020. See: https://www.weforum.org/agenda/2020/02/irreversible-emissions-permafrost-tipping-point
6. "All About Frozen Ground, Methane and Frozen Ground", Kevin Schaefer, *National Snow and Ice Data Center*, 2021. See: https://nsidc.org/cryosphere/frozenground/methane.html
7. *Climate Change 2014 Synthesis Report. Summary for Policymakers*, Intergovernmental Panel on Climate Change (IPCC), 2014.

8. For further information see the Great Barrier Reef Marine Park Authority's article on coral bleaching at: https://www.gbrmpa.gov.au/the-reef/reef-health/coral-bleaching-101

CHAPTER 17: NATURE RELIEVED OF ITS HUMAN BURDEN

1. "Chernobyl", via *History.com*, June 2019. See: www.history.com/topics/1980s/chernobyl

2. "30 years living with Chernobyl, 5 years living with Fukushima: Health effects of the nuclear disasters in Chernobyl and Fukushima", Dr Angelika Claußen and Dr Alex Rosen, *International Physicians for the Prevention of Nuclear War*, Germany, April 2016.

3. "Consequences of the radiation accident at the Mayak production association in 1957 (the 'Kyshtym Accident')", AV Akleyev et al., *Journal of Radiological Protection*, 37 R19, 2017. See: https://iopscience.iop.org/article/10.1088/1361-6498/aa7f8d

4. "Animals Rule Chernobyl Three Decades After Nuclear Disaster", John Wendle, *National Geographic*, April 2016. See: www.nationalgeographic.com/news/2016/04/060418-chernobyl-wildlife-thirty-year-anniversary-science

5. *The Ancestor's Tale: A Pilgrimage to the Dawn of Life*, Richard Dawkins, Weidenfeld & Nicolson: London, 2004.

6. *Trees That Call Australia Home,* John Halkett, Potts Point Publishing: Sydney, 2009.

7. "The last mammoths died on a remote island", *ScienceDaily* via University of Helsinki, 7 October 2019. See: https://www.sciencedaily.com/releases/2019/10/191007081750.htm

8. "Why Did New Zealand's Moas Go Extinct?", Virginia Morell, *AAAS Science*, March 2014. www.sciencemag.org/news/2014/03/why-did-new-zealands-moas-go-extinct

9. Dichlorodiphenyltrichloroethane, commonly known as DDT, is a colourless, tasteless, and almost odourless crystalline chemical compound originally developed as an insecticide. Now banned in many countries because of its adverse environmental impacts.

CHAPTER 18: ENTER THE CORONA VIRUS

1. "Covid19 is nature's wakeup call to complacent civilisation", George Monbiot, *The Guardian*, 25 March 2020.

2. "Corona virus is a dress rehearsal for what awaits us if governments continue to ignore science", John Hewson, *The Guardian*, 22 April 2020.

3. "Corona virus: Nature is sending us a message", Damian Carrington, *The Guardian*, 25 March 2020.

4. "Destruction of wildlife and the climate crisis is hurting humanity, with Covid-19 a 'clear warning shot', say experts." in "Corona virus: Nature is sending us a message", Damian Carrington, *The Guardian*, 25 March 2020.

5. "Scientific COVID-19 strategy must apply to climate change", *The Sydney Morning Herald*, Editorial, 7 May 2020.

6. "Jane Goodall says global disregard for nature brought on coronavirus pandemic", Kirsten Diprose and Matt Neal, *Australian Broadcasting Corporation*, 11 April 2020. See: https://www.abc.net.au/news/2020-04-11/jane-goodall-says-disregard-for-nature-has-brought-coronavirus/12142246

7. "Corona virus: Nature is sending us a message", Damian Carrington, *The Guardian*, 25 March 2020.

8. "The danger and cruelty of wet markets", Animal Equity United Kingdom, April 2019. See: https://animalequality.org.uk/act/ban-wet-markets

9. "Climate crisis: in coronavirus lockdown, nature bounces back – but for how long?", Jonathan Watts, *The Guardian*, 10 April 2020. See: https://www.theguardian.com/world/2020/apr/09/climate-crisis-amid-coronavirus-lockdown-nature-bounces-back-but-for-how-long

10. Ibid.

11. "Nurture to nature via COVID-19, a self-regenerating environmental strategy of environment in global context", *ScienceDirect*, Science of The Total Environment, 10 August 2020. See: https://www.sciencedirect.com/science/article/pii/S004896972032605X

12. "New study: Fewer animals killed by cars as COVID-19 keeps motorists off the road", Ethan Shaw. *Earth Touch News Network*, 30 June 2020. See: https://

www.earthtouchnews.com/conservation/
human-impact/new-study-fewer-
animals-killed-by-cars-as-covid-19-
keeps-motorists-off-the-road/

13. "Stop! French philosopher Latour urges
    no return to pre-lockdown normal",
    Benoit van Overstraeten. *Reuters*, 9 May
    2020.

CHAPTER 19: SLEEPWALKING PAST A POINT OF NO RETURN

1. UN Climate Change Conference COP
   25 (2–13 December 2019). See: https://
   unfccc.int/cop25

2. "Don't fiddle while the planet burns,
   U.N. chief warns climate summit.",
   Isla Binnie and Jake Spring, Reuters,
   2 December 2019. See: https://www.
   reuters.com/article/us-climate-change-
   accord-idUSKBN1Y60NS

3. Update, "Don't fiddle while the planet
   burns, U.N. chief warns climate
   summit", Isla Binnie and Jake Spring,
   Reuters, 2 December 2019. See:
   https://money.yahoo.com/1-u-n-chief-
   opens-115019707.html

4. *COP25: Key outcomes agreed at the UN
   climate talks in Madrid*, Simon Evans and
   Josh Gabbatiss, Carbon Brief, Climate
   Diplomacy, 16 December 2019. See:
   https://www.climate-diplomacy.org/
   news/cop25-key-outcomes-agreed-un-
   climate-talks-madrid

5. "Greta Thunberg condemns world leaders
   in emotional speech at UN", Oliver
   Milman, *The Guardian*, 24 September
   2019. See: https://www.theguardian.
   com/environment/2019/sep/23/greta-
   thunberg-speech-un-2019-address

6. *World scientists declare climate emergency*,
   Watts Up With That?, 6 November
   2019. See: https://wattsupwiththat.
   com/2019/11/06/world-scientists-
   declare-climate-emergency/

CHAPTER 20: TOWARDS THE CLIFF EDGE

1. Revelation 11:18, *The Catholic
   Comparative New Testament*, Oxford
   University Press. 2005.

2. "Tipping Points and Climate Change:
   Metaphor Between Science and
   the Media", Sandra van der Hel,
   Lina Hellsten and Gerard Steen,
   Environmental Communication, Vol 12,
   2018. See: www.tandfonline.com/doi/full
   /10.1080/17524032.2017.1410198

3. "Nature crisis: Humans 'threaten 1m
   species with extinction'", Matt McGrath,
   BBC Science News, 6 May 2019.

4. "The Climate Crisis Moment Has Now
   Come, Warns Sir David Attenborough",
   conserve Energy Future 2021. See:
   https://www.conserve-energy-future.
   com/climate-crisis-moment-has-now-
   come-warns-sir-david-attenborough.php

5. *A Life on Our Planet: My Witness
   Statement and a Vision for the Future*,
   David Attenborough, Witness Books:
   Pelican Random House, 2020, pp 95,
   100, 121.

6. Ibid. p 105.

7. Ibid. p 121.

8. Ibid. p 220.

BIBLIOGRAPHY
and suggested additional reading

Adams, Max, 2014, *The Wisdom of Trees*, Head of Zeus Ltd, London, United Kingdom.

Archer, Michael and Beale, Bob, 2004, *Going Native: living in the Australian environment* Hodder, Sydney.

Attenborough, David, 2020, *A Life on Our Planet: My Witness Statement and a Vision for the Future*, Witness Books, Pelican Random House.

Baskin, Yvonne, 1997, *The Work of Nature: How the diversity of life sustains us*, Island Press, Washington DC.

Brooker, Ian and Kleinig, David, 1983, *Field Guide to Eucalypts: Southern-eastern Australia*, Inkata Press, Melbourne.

Carey, Francis, 2012, *The Tree: Meaning and Myth*, British Museum Press, London.

Carson, Rachel, 1962, *Silent Spring*, Houghton Mifflin, New York.

Dargavel, John, 1995, *Fashioning Australia's Forests*, Oxford University Press, Melbourne.

Dawkins, Richard, 2004, *The Ancestor's Tale: A Pilgrimage to the Dawn of Life*, Weidenfeld & Nicolson, London.

Darwin, Charles, 1859, *On the Origin of Species by Means of Natural Selection, or the Preservation of Favoured Races in the Struggle for Life*, Royal Society, United Kingdom.

de Perthuis, Christian and Jouvet, Pierre-Andre, 2015, *Green Capital: In A New Perspective on Growth*, Columbia University Press, New York.

Fenwick, Steffen and Jacqui, 2016, *The Heat Marches On* 2016, Climate Council of Australia.

Flanagan, Richard, 2020, "Australia Is Committing Climate Suicide", *The New York Times*, New York.

Flannery, Tim, 2010, *Here on Earth: an argument for hope*, Text Publishing, Melbourne.

Flannery, Tim, 2005, *The Weather Makers: The History and Future Impact of Climate Change*, Text Publishing, Melbourne.

Flannery, Tim, 1998, *The Future Eaters: An Ecological History of the Australasian Lands and People*, Reed New Holland, Sydney.

Fortey, Richard, 2004, *The Earth: An Intimate History*, Harper Collins, London.

Garnaut, Ross, 2019, *Superpower: Australia's Low-Carbon Opportunity*, La Trobe University Press, Melbourne.

Gore, Al, 2007, *The Assault on Reason*, Bloomsbury Publishing, London.

Hageneder, Fred, 2005, *The Living Wisdom of Trees*, Duncan Baird, London.

Hall, Norman, Johnston, RD and Chippendale, GM, 1970, *Forest Trees of Australia*, Australian Government Publishing Service, Canberra.

Hay, Ashley, 2002, *Gum*, Duffy & Snellgrove, Sydney.

Helm, Dieter, 2016, *Natural Capital: Valuing the Planet*, Yale University Press, Connecticut.

Holmes, D, 2002, *Where Have All the Forests Gone?*, EASES Discussion Paper, World Bank, Washington DC.

Holliday, Ivan, 1989, *A Field Guide to Australian Trees*, Lansdowne Publishing, Sydney.

Huikari, Olavi, 2012, *The Miracle of Trees*, Walker Publishing, New York.

Lines, William J, 1991, *Taming the Great Southern Land: A History of the Conquest of Nature in Australia* Allen & Unwin, Sydney.

Mardas, N, Mitchell, A, Crosbie, L, Ripley, S, Howard, R, Elia, C and Trivedi, M, 2009, *Global Forest Footprints: How Businesses around the World Contribute to Deforestation – the Risks of Inaction and the Opportunity for Change*, Forest Footprint Disclosure Project, Global Canopy Programme, Oxford.

McCalman, Iain, 2010, *Darwin's Armada*, Pocket Books, London.

Norhaus, William, 2013, *The Climate Casino: Risk, Economics and Uncertainty for a Warming World*, Yale University Press, New Haven.

O'Grady, AP, Pinkard, EA, Mount, RE, Schmidt, RK, Cresswell, ID and Stewart SB, 2020, *Conceptual model to support natural capital accounting of a forestry enterprise A report from the lifting farm gate profits: the role of natural capital accounts project*, CSIRO Land and Water, Canberra.

Pakenham, Thomas, 2015, *The Company of Trees: A Year in a Lifetime's Quest*, Weidenfeld & Nicolson, London.

Pakenham, Thomas, 2002, *Remarkable Trees of the World*, Weidenfeld & Nicolson, London.

Pascoe, Bruce, 2014, *Dark Emu*, Magabala Books, Aboriginal Corporation, Broome.

Paspadakis, Elim, 1998, *Politics and the Environment: The Australian Experience*, Allen & Unwin, Sydney.

Prentice, IC, et al, 2001, *The Carbon Cycle and Atmospheric Carbon Dioxide*, www.grida.no/climate/ipcc_tar/wg1/pdf/tar-03.pdf

Roland, Ennos, 2001, *Trees*, Natural History Museum, London.

Simard, Suzanne, 2021, *Finding the Mother Tree*, Allen Lane, London.

Southwood, Richard, 2003, *The Story of Life*, Oxford University Press, Oxford.

Standish, Robert, 1960, *The First of Trees*, Phoenix House, London.

Suzuki, David, 2013, *The Legacy: An Elder's Vision for our Sustainable Future*, Allen & Unwin, Vancouver.

Taylor, Griffith, 2010, *Australia in its Physiographic and Economic Aspects*, Kessinger Publishing, Montana.

Tudge, Colin, 2000, *The Variety of Life: A Survey and a Celebration of All Creatures that Have Ever Lived*, Oxford University Press, Oxford.

Weisman, Alan, 2007, *The World Without Us*, St. Martin's, Thomas Dunne Books, New York.

Wilson, EO, 1992, *The Diversity of Life*, Harvard University Press, Massachusetts.

Wohllben, Peter, 2016, *The Hidden Life of Trees*, Black Inc, Victoria.

INDEX

ACKNOWLEDGEMENTS

Much of this book is based on the observations, research and writings of numerous biologists, foresters, climate change scientists and other researchers. This body of work is acknowledged with gratitude.

In Chapter 1 I acknowledge the credentials and contribution of John Hewson in supporting this book and for writing the preamble. Again, I thank John for his efforts and his intellect in arguing a cogent case for action on climate change.

I also acknowledge with appreciation the significant contribution that economist Kenneth Ng has made to Chapters 14 and 15 that deal with natural capital accounting and carbon pricing matters. Thank you Kenneth.

I also thank others who have helped to get this book into your hands. Notably Jan Hume provided great assistance with fine tuning the manuscript so it passed muster and Russell Jeffery of Emigraph Creative in Sydney provided welcomed help in preparing all the illustrations.

Critically the personal attention of Matthew Richardson of Halstead Press to the rigour of the text, and his team for their efforts in design and publishing of the book, are warmly recognised.

My grateful thanks and appreciation to you all.

ABOUT THE AUTHOR

John Halkett is the author of numerous scientific papers and books about the environment, and is a leader in developing industries that have trees, sustainable forest management, biomass and waste wood use at the centre of carbon emissions reduction, bioenergy and green hydrogen production in Australia's Hunter Valley. He runs a forest consultancy business and has expertise in temperate and tropical forest management and forest-based industries. He has worked in the USA, China, Canada, Chile, PNG, Myanmar, and Africa, and held senior positions in government forest and conservation agencies in Australia and New Zealand.

OTHER BOOKS BY JOHN HALKETT

The World of the Kauri (with E V Sale) Reed Methuen, Auckland, 1986.

The Native Forests of New Zealand GP Books, Wellington, 1991.

Tree People (with Peter Berg and Brian Mackrell) GP Books, Wellington, 1991.

Trees that Call Australia Home Potts Point Publishing, Sydney, NSW, 2009.

Jungle Jive: Sustaining the Forests of Southeast Asia, Connor Court Publishing, Brisbane, 2016.

By the Light of the Sun: Trees, Wood, Photosynthesis and Climate Change, Connor Court Publishing, Brisbane, 2018.